Home Wiring from Start to Finish

2nd Edition

Home Wiring from Start to Finish

2nd Edition

Robert W. Wood

TAB Books
Imprint of McGraw-Hill

New York San Francisco Washington, D.C. Auckland Bogotá
Caracas Lisbon London Madrid Mexico City Milan
Montreal New Delhi San Juan Singapore
Sydney Tokyo Toronto

NOTICES
National Electrical Code®   National Fire Protection Association
Romex®   General Cable Company

pbk   4 5 6 7 8 9 10 11 12 13   FGR/FGR   9 9 8 7 6 5
hc   1 2 3 4 5 6   7   8   9 10   FGR/FGR   9 9 8 7 6 5 4 3

Wood, Robert W.
   Home wiring from start to finish / by Robert W. Wood.—2nd ed.
     p.   cm.
   Includes index.
   ISBN 0-8306-4185-8   ISBN 0-8306-4186-6 (pbk.)
   1. Electric wiring, Interior—Amateurs' manuals.   I. Title.
TK3285.W67   1993
621.319'24—dc20   93-1538

Acquisitions Editor: Kimberly Tabor
Book Editor: April D. Nolan
Director of Production: Katherine G. Brown
Computer Artists: Ruth Gunnett
                          Toya Warner
Coding: Ollie Harmon
Layout: Rhonda Baker
             Wendy Small
             Donna Harlacher
             Jan Fisher
Proofreading: Nancy Mickley
Indexing: Stacey R. Spurlock   TAB1
Book Design: Jaclyn J. Boone   4254

Contents

Introduction

We seldom realize how much we depend on electricity—how the comfort and efficiency we enjoy in our lives would not be possible without this convenient form of energy. It is only when we experience a power outage that we realize how everything seems to come to a standstill. Supermarkets can't sell food because the checkout registers won't work. You can't buy gasoline for your car because the pumps won't work. The food in the refrigerators and freezers spoils. And your only source of news is a battery-operated radio or yesterday's newspaper. Fortunately, these blackouts seldom happen, and when they do, utility companies rush repair crews to the problem area and restore service quickly.

Our energy demands require that residential electrical systems be well planned to cover today's needs, as well as future requirements. The electrical system also must be correctly installed.

The use of electricity couldn't just happen. It had to evolve, sometimes painfully, to the safe standards we have today. In 1897, an electrical document was developed through the combined efforts of representatives from insurance, architectural, electrical, and associated groups. This document became the National Electric Code (NEC) and has been sponsored by the National Fire Protection Association since 1911. The NEC normally is referred to as simply "the Code."

Developed over many years, the Code is a set of rules that have been found to provide safe and practical electrical installations. The only purpose of the Code is to provide rules for the safe installation of electrical wiring. It is designed for the protection of life and property and does not have the force of law. Note, however, that most state and local governing bodies do adopt the Code and might even include additional rules for their particular areas.

The Code is updated and revised every three years. Copies are available at bookstores or from:

NATIONAL FIRE PROTECTON ASSOCIATION
Batterymarch Park
Quincy, MA 02269

A safely installed wiring system can be defeated if any devices or components are used that are manufactured below minimum safety standards. These standards were established by a nonprofit testing laboratory known as Underwriters' Laboratories Inc. (UL) or by the American National Standards Institute (ANSI). Manufacturers submit their product to the laboratory for testing. If the product meets minimum safety standards, the product is listed by the laboratory and will display the familiar UL label. Inspectors monitor production at the factories to ensure that the product continues to maintain safety requirements. The laboratory further tests samples bought in stores. If the product meets the minimum safety requirements, it will continue to be UL listed.

When a product displays the UL symbol, it has passed the minimum safety requirements if the device is used for the purpose for which it was intended. The UL symbol is not a reliable indicator of quality. The same product made by a number of manufacturers might all have the UL label, but one product of a higher quality might outlast a lower-quality product many times. Most stores today will only sell merchandise that is UL listed. The important thing is to shop for quality and to do quality work.

An apprentice electrician might spend four or more years learning the trade before being classified as a journeyman. Because of their training and experience, electricians are usually in the upper part of the pay scale. They work fast and efficiently.

Home Wiring from Start to Finish, 2nd ed., however, was written for the homeowner who has time to work at a slower pace and to refer to reference material when necessary. This book provides a basic understanding of electricity. With its guidance, any person slightly handy with tools will be able to handle most residential wiring jobs, such as installing an electrical system, adding special circuits, making installations in rural areas, and making electrical repairs. The book also stresses safety and the importance of coordinating any projects with the local building authority and utility company.

Home Wiring from Start to Finish, 2nd ed. was written for the homeowner/builder with the do-it-yourself spirit—the one who gains monetary- and self-satisfaction from the completion of quality work.

Chapter **1**

Electricity & electrical circuits

Electricity is a natural form of energy that is all around us. We can feel it in our bodies as static electricity, and we can see it in the sky as lightning. It has no weight, size, color, or odor. Our knowledge of electrical energy is based more on what it can do than on what it is.

In its basic form, electricity performs little service. It must be changed into some other type of energy to become useful. Electrical energy is changed into light because of the resistance of a filament in a light bulb. It is changed into mechanical energy when current is applied to an electrical motor. Some appliances, such as electric toasters and irons, are designed to convert electrical energy into heat.

Electricity is versatile. It provides us with light, controls the comfort in our homes, and powers most of our household appliances.

To understand the source of electrical energy, we must look at an atom. Atoms exist in all forms of matter. They are made up of particles called protons, neutrons, and electrons. We are interested in the particles called electrons.

Normally, electrons will remain in a single atom; however, some can, and do, move from one atom to another, becoming *free electrons*. Copper, steel, and aluminum are made up from atoms that have many free electrons; as a result, these metals are good *conductors* of electricity. Copper is the best of the three (FIG. 1-1). Other materials—such as rubber, plastic, paper, and wood—have few or no free electrons and, consequently, do not conduct electricity efficiently and are called *insulators*.

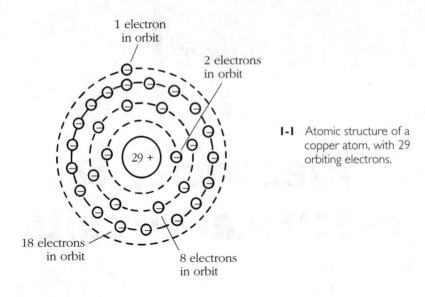

1-1 Atomic structure of a
copper atom, with 29
orbiting electrons.

1 electron
in orbit

2 electrons
in orbit

29 +

18 electrons
in orbit

8 electrons
in orbit

ELECTRICAL CURRENT

The electrical energy (or electromotive force) called *voltage* is supplied by
two types of electrical currents. The simplest of the two is generally provided by batteries and is called *direct current* or *dc*. The other type of current is *alternating current*, or *ac*; this is the type of current generated by
utility companies for home use (FIG. 1-2).

Direct current flowing in only one direction along a conductor or in a
circuit is a direct current. This type of current is normally provided by batteries but, in some cases, might be supplied by certain types of generators.
Direct current also can be produced by feeding alternating current into a
network of filters (capacitors and resistors) and rectifiers.

When a current flows first in one direction and then the other, the
current is said to be alternating. Current reversals might occur at any rate
from just a few per second up to hundreds, depending on the method of
generation. A standard was adopted in the early years of development,
and now the current flowing in homes and factories throughout the United States reverses itself 120 times each second. It takes two reversals to
make up one cycle (FIG. 1-3). All homes in the United States operate on
the same frequency: an alternating current of 60 cycles per second. The
unit of frequency that measures cycles per second is called *hertz (Hz)* in
honor of Heinrich Hertz, a German physicist of the late 1800s. A current
with a frequency of 60 cycles per second, then, is termed 60 Hz.

How current flows

When current flows through a wire, a magnetic field is built up around the
wire. Normally this field is very small, but if the wire is wound into several
coils, the magnetic field of one coil tends to add to the magnetic field in
the next coil, creating a strong magnetic field quickly. If a loop of wire

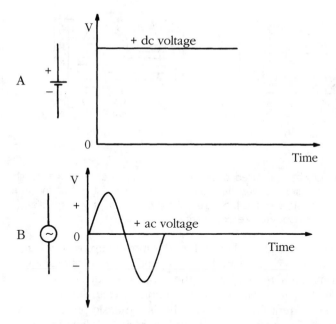

1-2 A dc voltage versus an ac voltage. (A) A dc voltage wave form shows a steady voltage with one polarity. (B) An ac wave form, often called a *sine wave*, shows the alternating voltage reversing in polarity.

1-3 The complete cycle of an ac wave form shows the constant changing in voltage value and polarity.

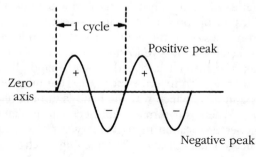

passes through the magnetic lines of force between the north and south poles of a magnet (FIG. 1-4), a small current can be detected in the wire. As the loop rotates, one side passes up through the lines of force while the other side of the loop moves down. The current in this first half-cycle flows in one direction.

When the loop arrives at the halfway position and neither side is going up or down, none of the lines of force are being cut and no electricity is generated. As the loop continues on into the second half of the cycle, the part of the loop that was formerly moving upward is now going down through the lines of force, and the side that was moving down is now going up. The current in the loop now flows in the opposite direction of the current induced in the first half of the cycle. When the loop again arrives at its

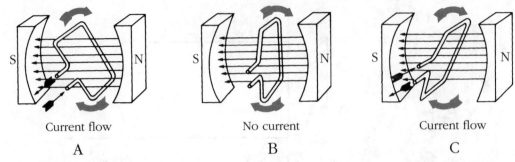

Current flow	No current	Current flow
A	B	C

1-4 Electricity being induced, or generated, in a wire loop moving through a magnetic field. (A) Current flows counterclockwise through the loop in the first half of the turn (left). (B) No current flows when the sides of the loop are not moving up or down (middle). (C) Finally, current flows again but in the opposite direction (right).

vertical position, with neither of the sides moving up or down, at that instant, none of the lines of force are being cut, and no electricity is being generated.

During each revolution of the loop, the current flows first in one direction then in the opposite direction. Twice during this revolution, there is no current flowing at all. This is the basic principle that produces alternating current commonly found in homes today. To make a simple generator practical, you would increase the number of wire loops and rotate them at a higher speed.

Electric utility companies operate huge generating stations that provide electrical power for cities and homes. This voltage is lowered in a series of steps through transformers until it reaches a final transformer located on power poles in residential neighborhoods. The service then continues on to the home through an underground or overhead service to a meter with an overload-protection device called the *main circuit breaker*. The meter has five dials with hands that rotate clockwise on every other dial. To read the meter, use the number the hand has passed, keeping in mind which way the hand is turning (FIG. 1-5).

When you turn on an electrical appliance, electrons flow from the generating station through the wires to the appliance and return back to the generating station. The moving electrons are called *current flow*. The amount of current flowing is measured in units called *amperes*, normally shortened to *amps*. The force that moves the electrons is an electromotive

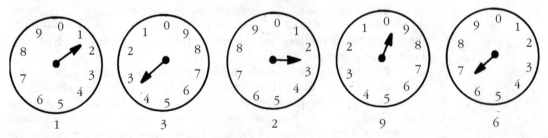

1-5 A meter reading of 13296. Last month's reading would be subtracted from this to determine the month's power consumption.

force measured in *volts*. The work performed by voltage and current is measured in *watts*. You can calculate these values by using two simple formulas called *Ohm's Law*. By covering one of the values in the circle in FIG. 1-6, the remaining will be the formula for determining that value. For example, to find the current when you know the watts of power and the voltage, cover the *I* in the circle and divide the power by the voltage.

Voltage ÷ Current = Resistance

Voltage × Current = Power in watts

1-6 Ohm's law.

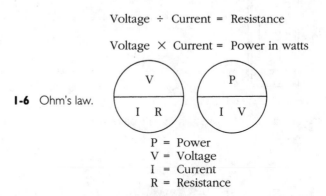

P = Power
V = Voltage
I = Current
R = Resistance

WHAT MAKES AN ELECTRICAL CIRCUIT?

To accomplish work, there must be a continuous path, or *circuit*, for the current to flow. (It might help you to visualize electrical circuits if you think of them as plumbing circuits and of the electron flow as water running in pipes.) Basically, an electrical circuit is simply a power source and a load (FIG. 1-7), with a switch or control device added (FIG. 1-8). The location of the switch is important. It is always installed in the hot (black) wire and not the return, neutral (white) wire. If the switch were installed in the neutral wire in FIG. 1-8, it would still turn the light on and off, but the light fixture would be live and, therefore, dangerous.

Each circuit must be protected by some overload device such as a fuse (FIG. 1-9) or a circuit breaker (FIG. 1-10). Note that the same size circuit

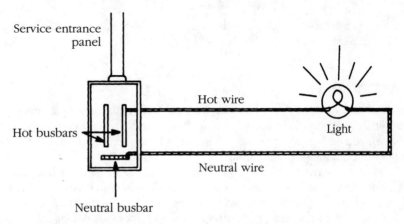

1-7 The completed circuit allows current to flow.

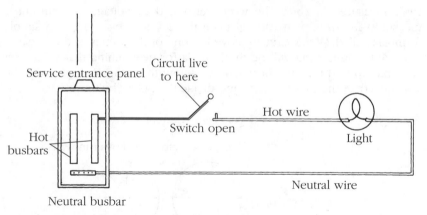

1-8 Most electrical circuits contain some type of switching device to control the flow of current.

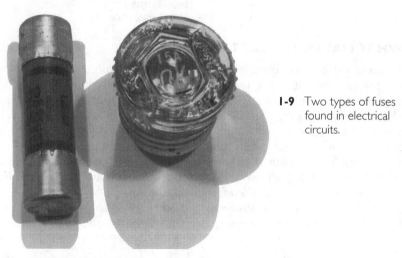

1-9 Two types of fuses found in electrical circuits.

breaker is available in different shapes. In addition, a 220-volt circuit must be protected by a double-pole breaker (FIG. 1-11).

SAFETY RULES

In electrical installations, safety has top priority. Materials must meet certain standards. Look for the UL stamp (FIG. 1-12) when purchasing materials, but remember it only means a minimum safety standard and does not reflect the quality of the material.

Most accidents are caused by poor work habits, so always keep safety in mind. Don't rush a project, and never work on a live circuit. Use ladders properly and keep the work area free of clutter to avoid falls.

One of the most important safety rules to remember is that the human body can conduct electricity. Because the substance that makes up our

1-10 Circuit breakers are more convenient than fuses because they only need to be reset instead of replaced.

1-11 A double-pole circuit breaker protects 240-volt and 120/240-volt circuits.

1-12 An Underwriters' Laboratory stamp means only that the product meets the minimum safety standards.

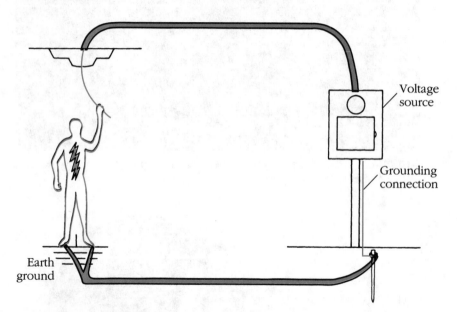

Voltage source

Grounding connection

Earth ground

1-13 Current will easily flow through a human body that completes a connection in an electrically live circuit.

planet is electrically neutral, current will flow when voltage is applied to a wire or any conductor connected to the ground. If this conductor happens to be the human body, severe shock will occur (FIG. 1-13). If a person receives a shock and is still part of the circuit, don't touch the victim with your bare hands. First shut off the power. If this is not possible, use some sort of insulating device, such as a coat or a broom, to remove the victim from the circuit. Then keep the victim warm and give artificial respiration, if necessary, until help arrives.

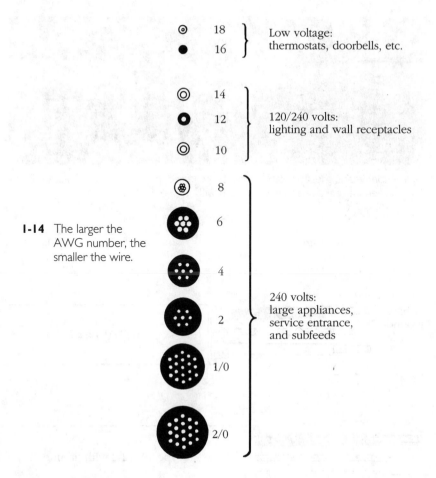

| | 18 | } | Low voltage: |
| | 16 | | thermostats, doorbells, etc. |

14
12 } 120/240 volts: lighting and wall receptacles
10

8
6
4
2
1/0
2/0

1-14 The larger the AWG number, the smaller the wire.

240 volts: large appliances, service entrance, and subfeeds

WIRE CAPACITY

The amount of current a wire can safely handle depends on the type of metal in the wire and its size. Copper wire is preferred over aluminum. The size of the wire is numbered using the American Wire Gauge (AWG) rating, where the smaller the number the larger the wire (FIG. 1-14). The larger the electrical load, the larger the wire must be. For example, a large load, such as an electric range, requires a #6 wire, while a small appliance, such as a lamp, is a much smaller load and requires only a #12 wire.

Nonmetallic sheathed cable, sometimes called Romex (the manufacturer's trade name), is widely used in residential wiring (FIG. 1-15). Cable is further classified for the conditions where it will be used, whether the location is wet or dry or buried in the ground (FIG. 1-16 and TABLE 1-1).

CIRCUITS

Each circuit begins at the service entrance panel, where it is protected by a circuit breaker. Each individual circuit is called a *branch circuit*

1-15 Nonmetallic sheathed cable is most often used in home wiring because it is relatively inexpensive and easy to work with.

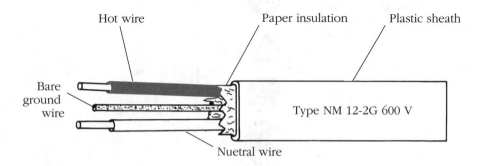

Hot wire   Paper insulation   Plastic sheath

Bare
ground
wire

Type NM 12-2G 600 V

Nuetral wire

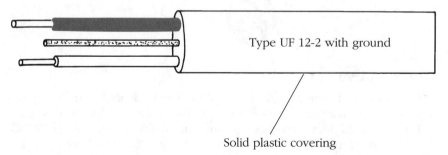

Type UF 12-2 with ground

Solid plastic covering

1-16 Cables are marked to indicate the wire size, the number of wires, and the type of location it might be used.

(FIG. 1-17). If a circuit feeds a subpanel that in turn feeds circuits in a shop or garage, it is called a *feeder circuit* (FIG. 1-18). The circuits that supply the lights and outlets in most areas of the home are called *general-purpose circuits*. These carry the smallest loads in the house and use #12 (sometimes #14) AWG cable.

The Code requires that at least two circuits (called *small-appliance circuits*) be provided to accommodate the receptacles in the kitchen, breakfast room, dining room, and pantry. Because today's kitchens have a

Table 1-1 Current-Carrying Capacities of Copper Wire

Ampacity of copper wire

Wire size	In conduit, cable, or buried directly in the earth		Single conductors in free air		
	Types T, TW	*Types RH, RHW, THW*	*Types T, TW*	*Types RH, RHW, THW*	*Weather-proof*
14	15*	15*	20*	20*	30
12	20*	20*	25*	25*	40
10	30*	30*	40	40*	55
8	40	50	60	70	70
6	55	65	80	95	100
4	70	85	105	125	130
2	95	115	140	170	175
1/0	125	150	195	230	235
2/0	145	2175	225	265	275
3/0	165	200	260	310	320

*Higher ampacities are given in the code book but the above figures are to be used for overcurrent protection.

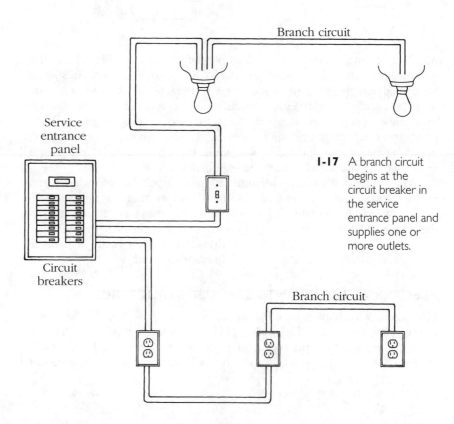

Branch circuit

Service entrance panel

Circuit breakers

Branch circuit

1-17 A branch circuit begins at the circuit breaker in the service entrance panel and supplies one or more outlets.

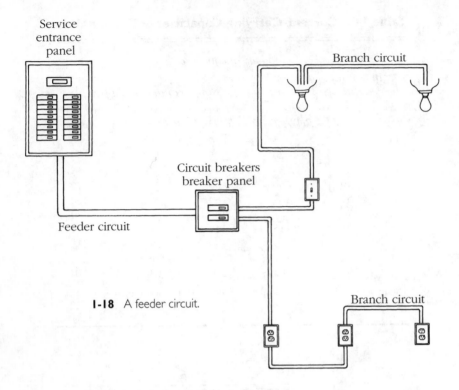

I-18 A feeder circuit.

number of small appliances—toasters, coffee makers, electric skillets, microwave ovens, and refrigerators (FIG. 1-19)—two 20-amp circuits should be considered the bare minimum. The receptacles above the countertop must be divided into at least two circuits. These circuits might also feed only these receptacles. Keep in mind that no lighting is connected to small-appliance circuits. Using an electric knife when the lights go out could be hazardous indeed. Also, while the refrigerator is not a small appliance, it is included in the small-appliance circuit.

The Code states that a separate 20-amp circuit be provided for the laundry receptacles. *Individual circuits* would include larger appliances, such as water heaters and ranges (FIG. 1-20). Such circuits run directly from the service entrance panel to only one outlet or appliance. Individual circuits include a circuit for a clothes dryer, central air/heat, dishwasher, garbage disposer, and any motor of ½ horsepower and over.

ELECTRICAL BLUEPRINTS & HOUSE STRUCTURE

Electricians wire houses based on a wiring plan drawn on the floor plans of the home (FIG. 1-21). The wiring plan just tells you where to locate the electrical equipment and the ceiling outlets, switches, and receptacles. How to run the wire or divide the circuits is determined on an individual

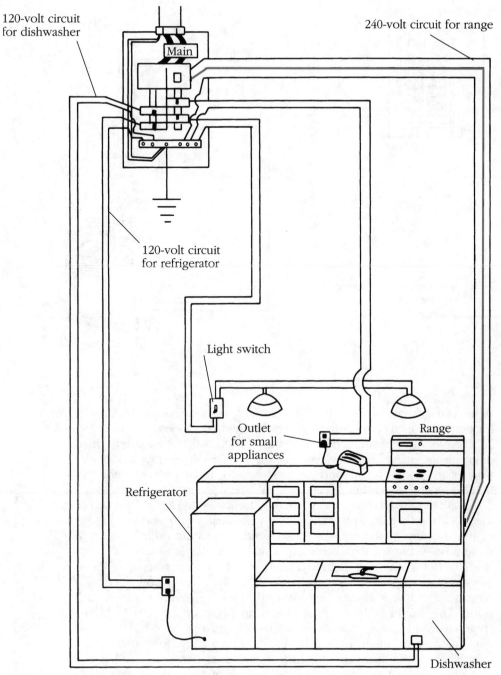

120-volt circuit
for dishwasher

240-volt circuit for range

Main

120-volt circuit
for refrigerator

Light switch

Outlet
for small
appliances

Range

Refrigerator

Dishwasher

1-19 Household circuits consist of several different electrical runs, but the kitchen area normally consumes the most energy.

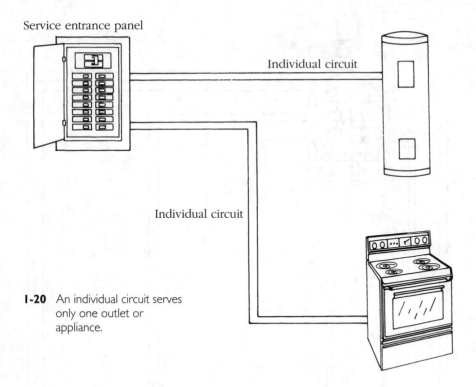

Service entrance panel

Individual circuit

Individual circuit

1-20 An individual circuit serves only one outlet or appliance.

basis. Often, homeowners will make changes to suit a particular lifestyle; for example, ceiling fans or indirect lighting might be added.

The location of the service entrance panel is determined by which direction the utility company provides the service—from the back or the front of the house, for instance. As you can see in FIG. 1-21, symbols are used to represent switches, receptacles, and ceiling outlets.

You must study wiring plans carefully to get a good mental picture of the job before beginning the work. You also must have some knowledge of the structure of the house. Houses tend to be built pretty much the same way—very expensive homes are usually only larger (FIG. 1-22).

The floor can be a poured concrete slab or constructed of floor joists. These are usually 2-x-6-inch or 2-x-10-inch lumber on edge running horizontally. If floor joists are used, they are covered with a subfloor. This might be 4-x-8-foot sheets of plywood or particleboard (pressed wood chips). The walls are usually made of 2-x-4-inch studs, but outside walls might have 2-x-6-inch studs to allow for more insulation. These studs stand on a board of the same dimension called the sole plate, and a top plate is fixed to the top of the stud. Ceiling joists made of 2-x-6-inch or 2-x-10-inch lumber rest on the top plate where the rafters are mounted.

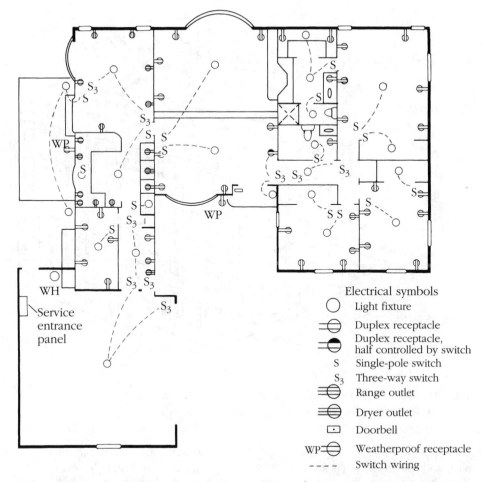

Electrical symbols

○ Light fixture

⊖ Duplex receptacle

⊖ Duplex receptacle, half controlled by switch

S Single-pole switch

S_3 Three-way switch

⊜ Range outlet

⊜ Dryer outlet

▭ Doorbell

WP⊖ Weatherproof receptacle

- - - - Switch wiring

1-21 A typical electrical wiring plan uses symbols to locate the switches and outlets.

TOOLS OF THE TRADE

When installing electric wiring, you will need some basic tools that are not normally found in the home (FIG. 1-23). Because you must know you are not working on a live circuit, almost any electrical work requires some type of voltage tester (FIG. 1-24). You also need a tape measure and wire strippers to quickly remove the insulation from the wire.

Two types of screwdrivers (Phillips and standard, flat-blade) will handle most of the screws. It is important to keep the ends of flat-blade screwdrivers square.

Use tongue-and-groove pliers (often called channel-lock or water-pump pliers) to tighten locknuts and cable connectors. If you use some type of solderless connectors, you will need a crimping tool. Long-nose or needlenose

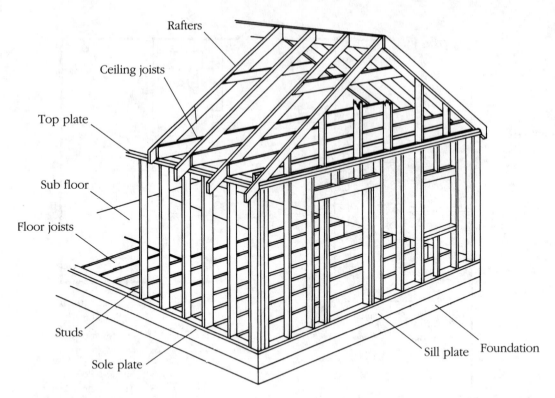

Rafters

Ceiling joists

Top plate

Sub floor

Floor joists

Studs

Sole plate

Sill plate   Foundation

1-22 The structure of a house. The electrical cable can be attached to or run through holes in the studs and floor and ceiling joists.

pliers are useful in bending wire loops to make connections at screw terminals, and *diagonal cutters* are handy for cutting wire ends in boxes or other close-fitting places (FIG. 1-25. *Lineman's pliers* have flat jaws that will pull and twist wires, while the side jaws will cut wires up to the larger sizes (FIG. 1-26). A small adjustable wrench will probably be helpful too, and an electric drill and wood bit will save time when drilling holes for the cable.

You will need a hammer to nail on boxes and to drive staples, and a wood chisel to cut out sawed notches and trim up cuts. A hacksaw is useful for cutting large cable or conductors. Use a level to level circuit breaker panels and to horizontally level and vertically plumb the positions for conduit. If you need to install conduit, you will need a tube bender and fish tape (FIG. 1-27). You can cut conduit with a hacksaw or a pipe cutter, but in either case, you should file smooth or ream out the cut end to prevent the insulation on the wires from being damaged.

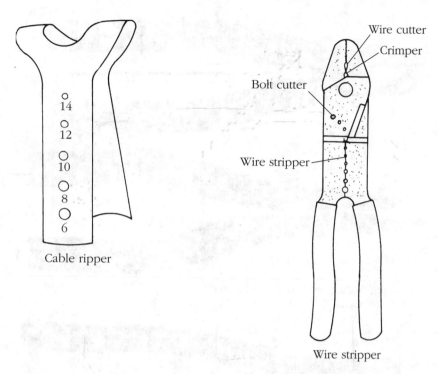

14

12

10

8

6

Cable ripper

Wire cutter

Crimper

Bolt cutter

Wire stripper

Wire stripper

1-23 Some tools are suitable only for electrical work.

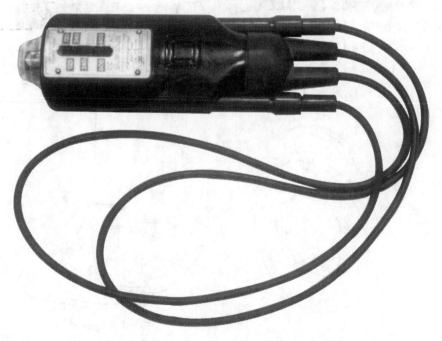

1-24 A rugged voltage tester is vital in electrical work.

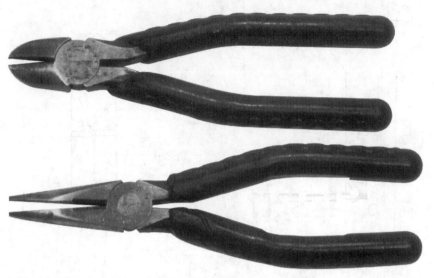

I-25 Needle-nose pliers and diagonal cutters.

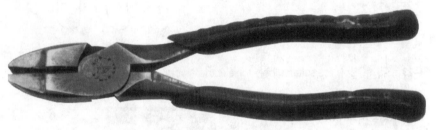

I-26 Lineman's pliers.

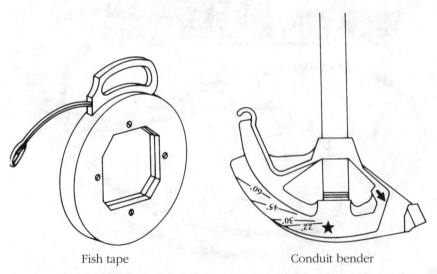

Fish tape   Conduit bender

I-27 Use a fish tape and conduit bender when wiring is to be installed in conduit.

Chapter **2**

Planning stage

When you are planning an electrical installation, give a lot of thought to future requirements. How will this service supply your needs 5, 10, or even 15 years from now? If you skimp on the original installation, you will find that adding outlets later will cost several times as much. By following the Code, you will have a safe installation, but it might not be a convenient or practical installation for your individual situation. You can achieve this only through careful planning.

Install a large enough service entrance. A 100-amp service might have been adequate a few years ago, but if you consider future needs, a 200-amp one might be necessary. Install plenty of receptacles. Avoid the use of pull-chains on light fixtures; install a switch instead. Three-way switches are a little more expensive, but the added convenience over the years will more than make up the difference.

In most cases, you will need a permit to do the work. Cooperation with the local building authority will make the project go much smoother. Always consult with the building inspectors before beginning any work; it might need to be done by a licensed electrician. Normally the utility company will not provide power until an inspection certificate has been turned in.

SERVICE ENTRANCE PANEL

When locating the service entrance panel, determine where the electric service will enter the house. Will it come from the back or the street side? Locate the service entrance panel as close to the service drop or meter as possible (FIG. 2-1).

If it is not convenient to install the panel next to the meter, you can install a main circuit breaker next to the meter instead. Wire the service to

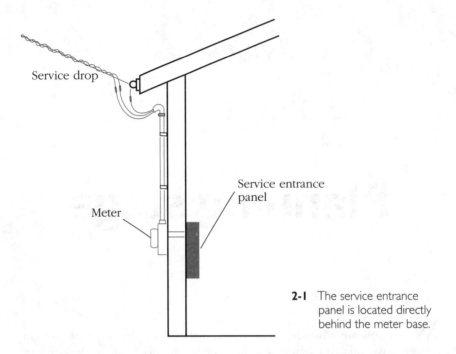

Service drop

Service entrance
panel

Meter

2-1 The service entrance
panel is located directly
behind the meter base.

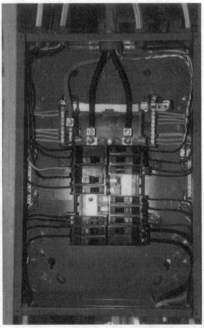

2-2 Installed next to the meter, the main
circuit breaker shows the wires feeding the service entrance panel.

2-3 The service entrance panel is fed
from the main breaker next to the
meter.

the meter base, then run it over to the main circuit breaker (FIG. 2-2). From there, run the wires to the service entrance panel in the desired location. In FIG. 2-3, there are only the circuit breakers for the branch circuits. The main breaker is over by the meter.

It is important both to locate the service entrance panel close to the area that uses the most power and to have the panel accessible at all times (FIG. 2-4). Service panels located on outside walls allow easy access for the ground wire to run down and through the wall to the ground rod (FIG. 2-5). You must locate the service entrance panel at a convenient height above the floor, with the bottom of the panel about 40 to 48 inches (3½ to 4 feet) above floor level. Use a crayon or a heavy marker to mark the location of the service entrance panel.

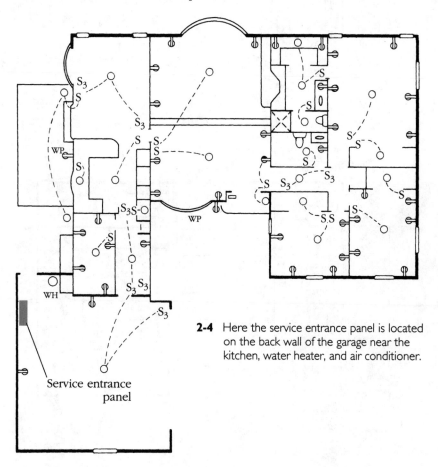

2-4 Here the service entrance panel is located on the back wall of the garage near the kitchen, water heater, and air conditioner.

GENERAL-PURPOSE CIRCUITS

The general-purpose circuits are the ones that supply the lighting outlets and most of the receptacles that are used for TVs, radios, reading lamps, and portable vacuum cleaners. Figure 2-6 is only one example of the several different ways to divide these circuits.

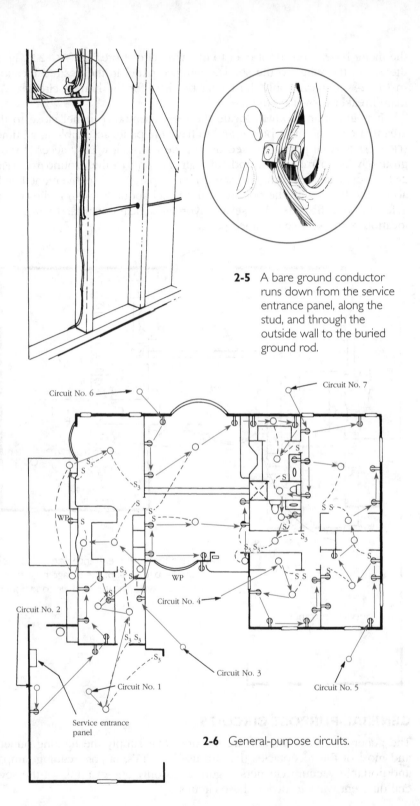

2-5 A bare ground conductor runs down from the service entrance panel, along the stud, and through the outside wall to the buried ground rod.

Circuit No. 6

Circuit No. 7

Circuit No. 2

Circuit No. 4

Service entrance panel

Circuit No. 1

Circuit No. 3

Circuit No. 5

WP

WP

2-6 General-purpose circuits.

SAMPLE CIRCUIT LAYOUT

Each circuit normally begins at the outlet closest to the service entrance panel. Usually #12 wire is considered to be the minimum size wire. In some cases, however, #14 is used (TABLE 2-1). About 12 outlets or less is a safe number for circuits using #12 wire. Number-14 wire can usually handle about 10 outlets.

Table 2-1   AWG Wire Size and Circuit Breaker Size for Residential Circuits

Type of circuit	AWG wire size (copper)	Maximum breaker size
General purpose	14	15
	12	20
Small appliance	12	20
Individual appliance	10	30
	6	50

Circuit #1

The general-purpose circuit shown in FIG. 2-7 might include the lighting outlets in the garage, laundry room, utility room, kitchen, and dining room, as well as two patio lights and one outdoor receptacle. From the service entrance panel, mark the location of the ceiling outlet in the garage, then the remaining outlets on the circuit. Remember to mark the three-way switches in the garage, laundry room, and dining room.

Locate and mark each switch and receptacle. Mark the studs, using the proper symbols. Measure and mark the height. For locations with two switches, just use two Ss. Mark switch boxes for mounting 48 inches from the floor to the center of the box and on the opposite side of where the door hinges will be (FIG. 2-8). Boxes for receptacles are mounted on studs, with the bottom of the box 12 to 14 inches from the floor (FIG. 2-9). (Some people use the length of their hammer handle as a guide.) Mark the location for the receptacles.

Circuit #2

The circuit shown in FIG. 2-10 could serve the two receptacles in the garage. If the utility room is to be a workroom where power tools will be used, this circuit also could serve the five receptacles in FIG. 2-13. Mark the outlets for this circuit.

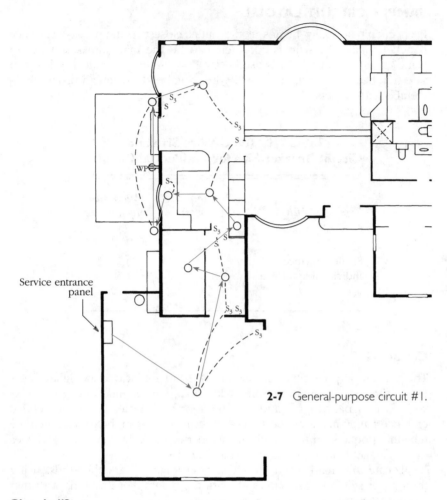

2-7 General-purpose circuit #1.

Circuit #3

Figure 2-11 illustrates a circuit that runs from the service entrance panel to the receptacle in the laundry room (not the washer or dryer). The circuit then runs to the receptacle and ceiling outlets in the living room. Mark the outlets for this circuit.

Circuit #4

The circuit in this illustration (FIG. 2-12) can feed the ceiling outlets and receptacles in one bedroom and two of the receptacles in another. Locate and mark these outlets.

Circuit #5

The next circuit can be used for the ceiling outlets, the remaining receptacles, and two of the receptacles in the master bedroom (FIG. 2-13). Mark the locations for these outlets.

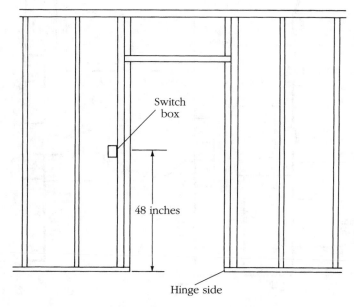

2-8 Standard location for switch boxes.

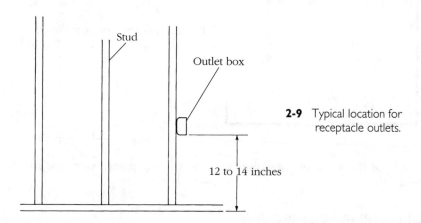

2-9 Typical location for receptacle outlets.

Circuit #6

The circuit shown in FIG. 2-14 can feed the receptacles and ceiling light in the family room as well as two lights and two receptacles in the master bath. Mark the outlets for this circuit.

Circuit #7

The circuit illustrated in FIG. 2-15 will supply the remaining receptacles and ceiling outlet in the master bedroom, one light in the master bath, a light and receptacle in the hall bathroom, and the hall light. Mark the

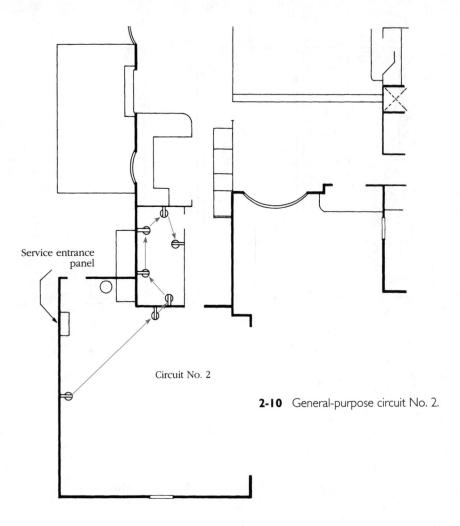

Service entrance
panel

Circuit No. 2

2-10 General-purpose circuit No. 2.

locations for these outlets, keeping in mind the three-way switches for
the hall light.

SMALL-APPLIANCE CIRCUITS

Small-appliance circuits are very important ones, so try to anticipate
your maximum needs and divide the circuits accordingly. In the sample
plan in FIG. 2-16, two circuits serve the receptacle outlets in the kitchen
and dining room. Appliance circuit #1 feeds part of the countertop,
while appliance circuit #2 feeds the rest of the countertop and the outlets in the dining room. The outlet for the refrigerator is included in
appliance circuit #1, but some prefer to run a separate circuit for this
appliance. Appliance circuit #3 is run for the clothes washer. Locate and
mark these outlets.

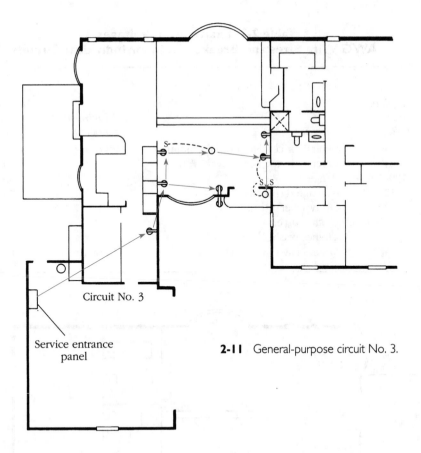

Circuit No. 3

Service entrance
panel

2-11 General-purpose circuit No. 3.

INDIVIDUAL CIRCUITS

Individual circuits serve the larger appliances and equipment such as the dryer, range, and water heater. The Code states that each must be on its own circuit.

For an individual circuit, a separate cable is run from the service entrance panel to the outlet serving the larger appliance or equipment (FIG. 2-17). It can be wired directly to the unit (range, water heater, etc.), or it can serve a receptacle to be used with a plug (for a clothes dryer, for instance).

The general-purpose and small-appliance circuits are wired for 120 volts. Some of the outlets on the individual circuits will require 240 volts (two hot wires) for a water heater, while a range will use a combination 120/240 volts (two hot wires and a neutral). The 120-volt circuits are usually wired with #12 wire and are protected by a 20-amp breaker. Wire and circuit breaker sizes for the 240-volt and 120/240-volt circuits will depend on the current requirements of the individual appliance (TABLE 2-2). Some local codes allow the dishwasher and garbage disposal to be on the same circuit. Check with your building inspector.

Table 2-2 Examples of Voltages, AWG Wire Sizes and Breaker Sizes for Individual Circuits

Typical size voltages	Circuit	AWG wire size (copper)	Maximum breaker (amps)
240	Central air condition or heat pump	8 or 6	40–60
240	Central electric heat	6	50
120/240	Range	10	30
120	Garbage disposer	12	20
120	Dishwasher	12	20
120/240	Clothes dryer	10	30
120	Clothes washer	12	20
240	Water heater	12 or 10	20–30

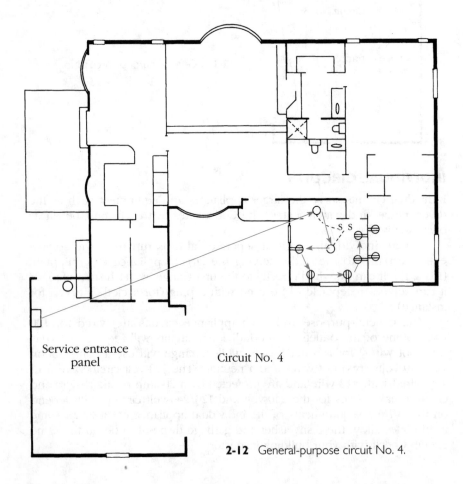

Service entrance panel

Circuit No. 4

2-12 General-purpose circuit No. 4.

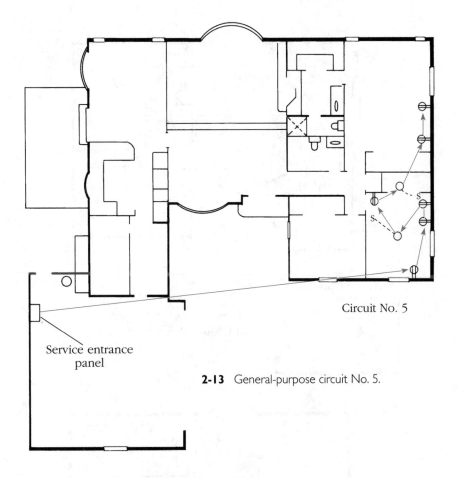

Circuit No. 5

Service entrance
panel

2-13 General-purpose circuit No. 5.

Mark these locations, identifying the 240-volt and 120/240-volt outlets. Each one requires a different size and type of outlet.

LOCATING GROUND-FAULT CIRCUIT INTERRUPTERS

Ground fault circuit interrupters (GFCIs) are amazing safety devices. They measure the current going in and out of a circuit or receptacle, which should be equal. Any variation indicates some of the current is going where it's not supposed to go and is creating a shock hazard. This is the ground fault.

The GFCI will interrupt a current leakage of as little as 4 to 6 milliamps (one-thousandths of an amp) in about $\frac{1}{40}$ of a second. This is much lower than the current that would trip a normal breaker. The easiest way to think of GFCIs is to remember that normal circuit breakers protect property while GFCIs protect people.

The Code requires GFCI protection on all 15-amp and 20-amp outdoor, bathroom, and some garage receptacles, as well as on temporary

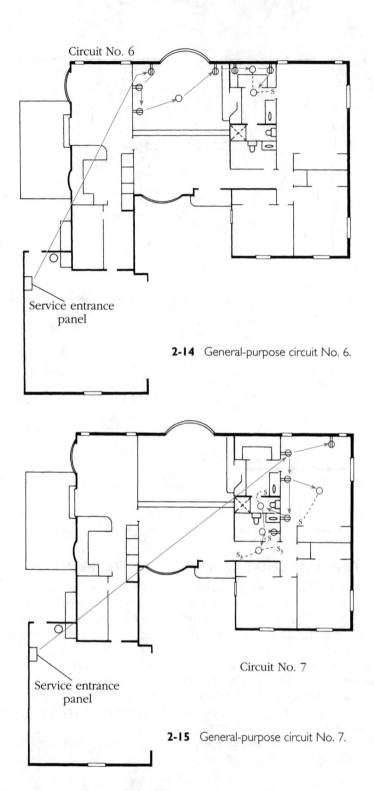

Circuit No. 6

Service entrance
panel

2-14 General-purpose circuit No. 6.

Circuit No. 7

Service entrance
panel

2-15 General-purpose circuit No. 7.

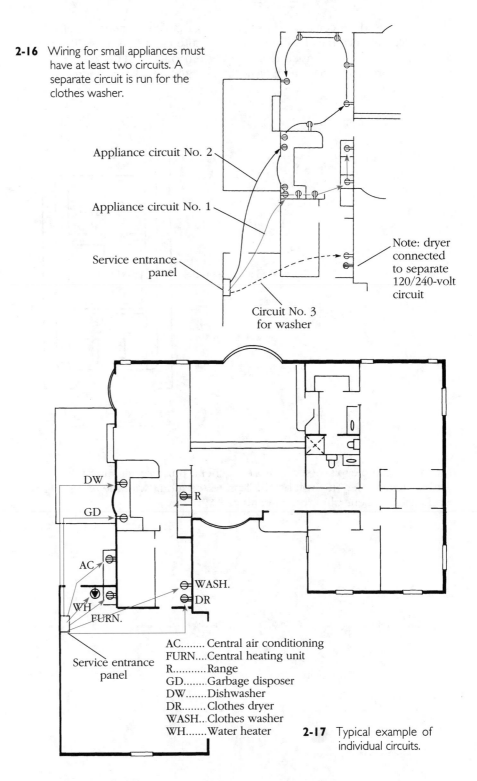

2-16 Wiring for small appliances must have at least two circuits. A separate circuit is run for the clothes washer.

Appliance circuit No. 2

Appliance circuit No. 1

Service entrance panel

Circuit No. 3 for washer

Note: dryer connected to separate 120/240-volt circuit

DW

GD

R

AC

WASH.

DR

WH

FURN.

Service entrance panel

AC........Central air conditioning
FURN....Central heating unit
R...........Range
GD........Garbage disposer
DW.......Dishwasher
DR........Clothes dryer
WASH...Clothes washer
WH.......Water heater

2-17 Typical example of individual circuits.

receptacles used on construction sites. A GFCI circuit breaker installed in the service entrance panel will protect the entire circuit, or a GFCI receptacle (FIG. 2-18) can be installed in the desired outlet. GFCIs cost a little more but, considering the safety factor, they might be a bargain indeed.

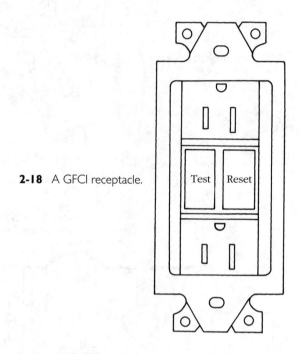

2-18 A GFCI receptacle.

In the sample plan (see FIG. 2-6), general-purpose circuit #1 has an outdoor receptacle. This will be GFCI-protected along with the two garage receptacles on general-purpose circuit #2.

Chapter **3**

Service entrance panel & boxes

The utility company delivers power to your home either through an overhead service or an underground service. Which service you receive might depend on local requirements.

The overhead service, or *service drop*, consists of wires that run from the power pole to the residence (FIG. 3-1). The wires must be located a minimum distance above ground, depending on the type of traffic (FIG. 3-2). Most service drops are made using *triplex cable*, or two insulated cables supported by a bare neutral cable (FIG. 3-3). After the house is completely wired and inspected, the utility company connects the service drop to the cable ends extending from the entrance head (FIG. 3-4).

Underground service is becoming more popular—even required in some subdivisions—because it requires little maintenance and it presents no potential danger from falling branches or high winds. Underground service can come from a *pad-mounted transformer* (FIG. 3-5) or from the utility company's power pole (FIG. 3-6). It is routed through conduit up to the meter (FIG. 3-7). The service must be buried to a minimum depth, depending on the type of installation (TABLE 3-1).

SERVICE SIZE & INSTALLATION

It is important to select a service large enough for both the needs you have now and any needs you might have in the future (FIG. 3-8). A 100-amp service is considered the minimum requirement for a residential service. If your home has electric heat or air-conditioning and an electric water heater, a 150-amp service might be necessary. To allow for future expansion, a 200-amp service is desirable. The utility companies usually will be glad to advise you on the size of service entrance equipment.

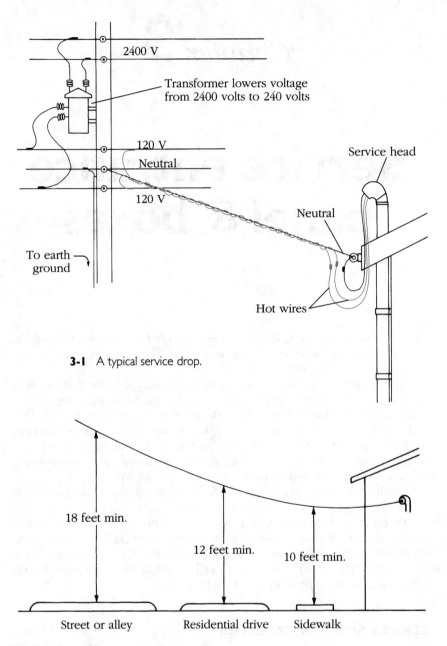

2400 V

Transformer lowers voltage
from 2400 volts to 240 volts

120 V

Neutral

120 V

Service head

Neutral

To earth
ground

Hot wires

3-1 A typical service drop.

18 feet min.

12 feet min.

10 feet min.

Street or alley

Residential drive

Sidewalk

3-2 Proper clearance above ground level must be maintained.

Installing the service & meter base

The size of the service entrance cable depends on the size of the service and if the cable is copper or aluminum (TABLE 3-2). Aluminum cable is less expensive than copper, but you must use a larger-size aluminum cable (FIG. 3-9).

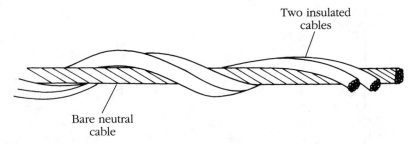

Two insulated
cables

Bare neutral
cable

3-3 The bare neutral cable serves as a support for the two insulated (hot) cables.

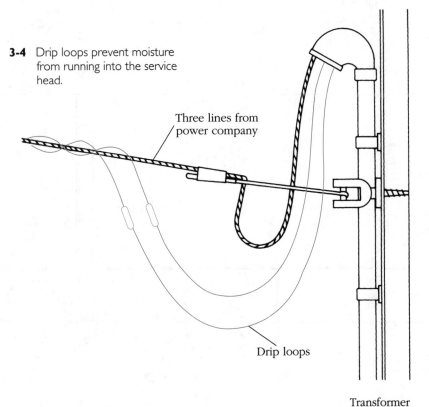

3-4 Drip loops prevent moisture
from running into the service
head.

Three lines from
power company

Drip loops

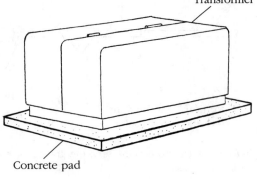

Transformer

3-5 A pad-mounted
transformer feeds an
underground service.

Concrete pad

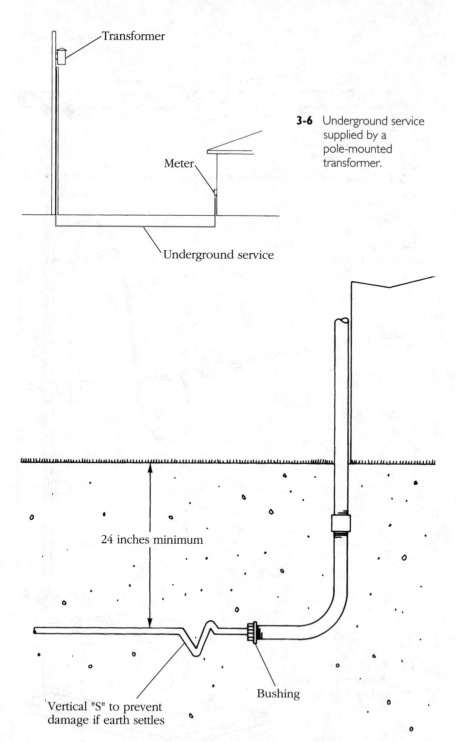

Transformer

Meter

Underground service

3-6 Underground service supplied by a pole-mounted transformer.

24 inches minimum

Vertical "S" to prevent damage if earth settles

Bushing

3-7 A bushing protects the service entrance cable where it enters the conduit at the meter base.

Table 3-1 Minimum Cover Requirements, 0 to 600 Volts

Wiring method	Minimum burial in inches
Direct buried cables	24
Rigid metal conduit	6
Intermediate metal conduit	6
Rigid nonmetallic conduit approved for direct burial	18

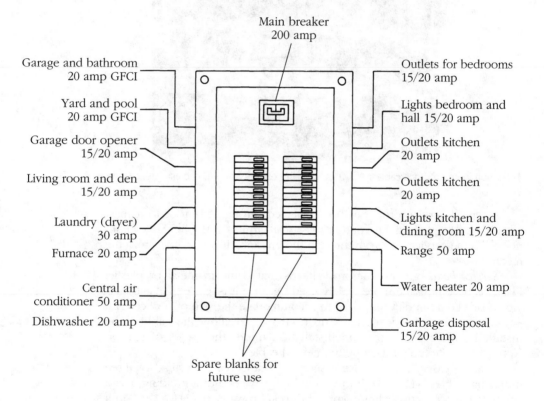

Main breaker
200 amp

Garage and bathroom
20 amp GFCI

Yard and pool
20 amp GFCI

Garage door opener
15/20 amp

Living room and den
15/20 amp

Laundry (dryer)
30 amp

Furnace 20 amp

Central air
conditioner 50 amp

Dishwasher 20 amp

Outlets for bedrooms
15/20 amp

Lights bedroom and
hall 15/20 amp

Outlets kitchen
20 amp

Outlets kitchen
20 amp

Lights kitchen and
dining room 15/20 amp

Range 50 amp

Water heater 20 amp

Garbage disposal
15/20 amp

Spare blanks for
future use

3-8 A well-planned service will include room for future expansion.

Table 3-2 Examples of Size of Services and Conductors

Size of service in amps	Copper AWG	Aluminum AWG
100	4	2
150	2	1/0
200	2/0	4/0

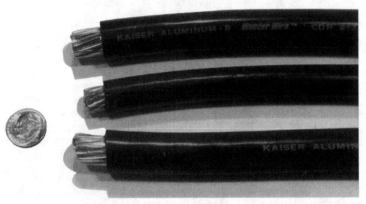

3-9 Aluminum entrance cables: A 2/0 cable in the center with 4/0 cables on each side.

A typical service drop can be connected through the roof to a mast usually made up of 2-inch conduit (FIG. 3-10). In FIG. 3-11, the cables connected to the service drop run through the entrance head and down to the meter base.

Service heads are weatherproof heads that come in metal or plastic. Plastic is becoming the more widely used because it is less expensive. Service heads come in different sizes to accommodate the size of service (FIG. 3-12). The top of the head snaps up (FIG. 3-13) to allow the cable to be installed. Figure 3-14 shows an installation under the eaves of a roof. Notice the white tape indicating the neutral cable.

In some cases, you can use a service entrance cable instead of three individual cables (FIG. 3-15). If the cable brings wires into a building, use a sill plate packed with sealing compound to keep water out (FIG. 3-16). In a service entrance cable (FIG. 3-17), the bare neutral wire is made up of a number of fine wires wrapped around the insulated wires. Twist these wires together to form a bare neutral wire (FIG. 3-18).

Position the meter base so its center is about 4 or 5 feet above ground level (FIG. 3-19). If you use a service entrance cable, feed it through a watertight connector at the top of the meter base; however, you can use a regular connector at the bottom (FIG. 3-20).

Support the service entrance cable at least every 4½ feet and within 12 inches of the entrance head, meter base, or any place it makes a connection (FIG. 3-21). When you use rigid conduit, screw it into a threaded hub (FIG. 3-22). Underground services often use PVC conduit feeding into the bottom of the meter base.

To wire the meter base:

1. Connect the two hot wires from the service entrance to the upper two terminals
2. Connect the two hot wires going to the service entrance panel to the bottom two terminals
3. Connect the neutral wire to the center neutral terminal (FIG. 3-23). Some utility companies prefer the neutral conductor not to be cut here (FIG. 3-24).

When making connections with aluminum cable, always coat the bare end with an antioxidant compound such as Penterox A13 or Alnox

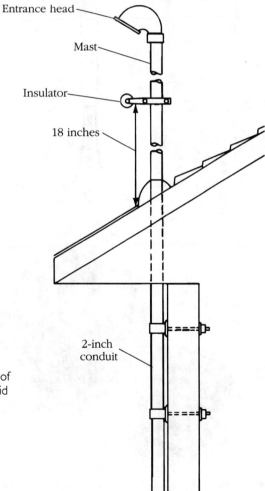

3-10 The mast through the roof overhang uses 2-inch rigid conduit.

(FIG. 3-25). When making any connection, cut away only enough insulation to allow the bare end to fit neatly in the terminals (FIG. 3-26) and make sure the connections are tight. Figure 3-27 shows a meter base wired and waiting for the meter to be installed.

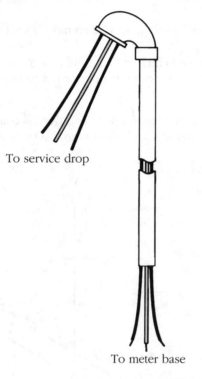

To service drop

To meter base

3-11 Individual cables run from the service drop to the meter base.

3-13 The top of the entrance head easily snaps in place.

3-12 Two sizes of plastic entrance heads.

3-14 An installation showing the neutral cable identified with white tape.

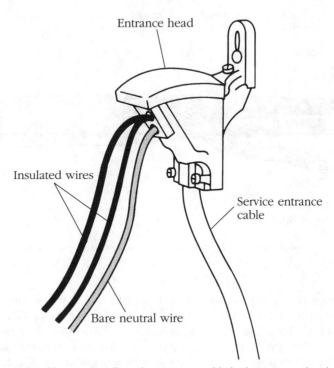

Entrance head

Insulated wires

Service entrance
cable

Bare neutral wire

3-15 You can install service entrance cable in the entrance head.

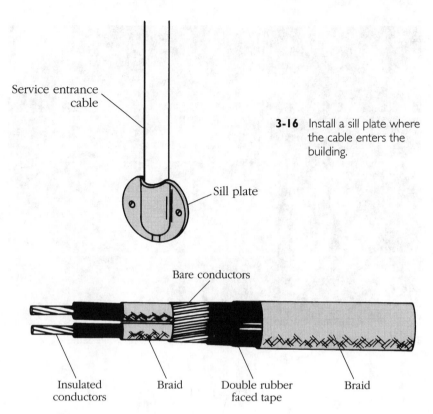

Service entrance
cable

3-16 Install a sill plate where
the cable enters the
building.

Sill plate

Bare conductors

Insulated
conductors

Braid

Double rubber
faced tape

Braid

3-17 Service entrance cable.

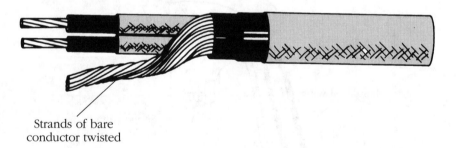

Strands of bare
conductor twisted

3-18 Service entrance cable with the bare wires twisted together to make the neutral
wire.

Service entrance cable is stiff and hard to bend. Where these heavy
cables are run in conduit and enter a building, an *entrance ell* is usually
required (FIG. 3-28). The ells come in metal or plastic and in different sizes
(FIG. 3-29). The cover comes with a gasket to protect it from moisture (FIG.
3-30). You will notice, also, that the cover includes both the UL stamp and
the size of the conduit connections.

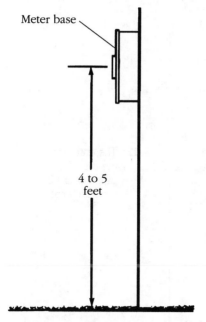

Meter base

4 to 5
feet

3-19 Locating the meter base above
ground level.

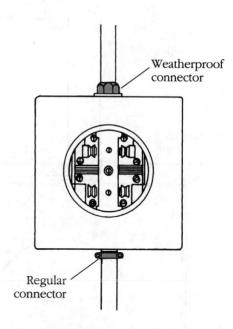

Weatherproof
connector

Regular
connector

3-20 Use a watertight connector at the top
of the meter base.

3-21 The cable must be
supported within 12
inches of cabinet and
every 4½ feet
elsewhere.

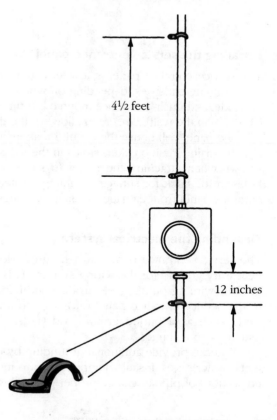

4½ feet

12 inches

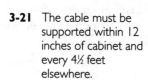

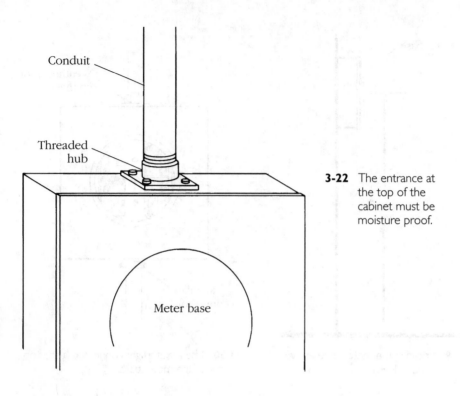

Conduit

Threaded
hub

Meter base

3-22 The entrance at the top of the cabinet must be moisture proof.

Installing the service entrance panel

The service entrance panel is a steel cabinet with a hinged door. Panels come in different styles, depending on whether they will be located outside or inside, and flush- or surface-mounted. If the panel is to be flush-mounted (recessed in the wall), be sure to allow for the thickness of the finished wall.

The panel will house the circuit breakers, the neutral busbar, and usually the main circuit breaker. Holes in the sides allow you to use mounting screws when attaching the panel to studs for flush mounting. Use the holes in the back for surface mounting. Protect wires entering the service entrance panel from damage by using bushings (FIG. 3-31).

Grounding the electrical system

The service entrance panel must be grounded (FIG. 3-32) to provide the necessary ground for the wiring system. This means you must connect a ground wire (grounding electrode conductor) from the neutral busbar in the service entrance panel (FIG. 3-33) to an approved grounding electrode such as a buried ground rod (FIG. 3-34). The grounding wire is usually a #6 or larger.

You can provide additional grounding by connecting a ground wire to metal water pipes. Install a jumper wire around the water meter or insulated sections of pipe to ensure proper grounding (FIG. 3-35).

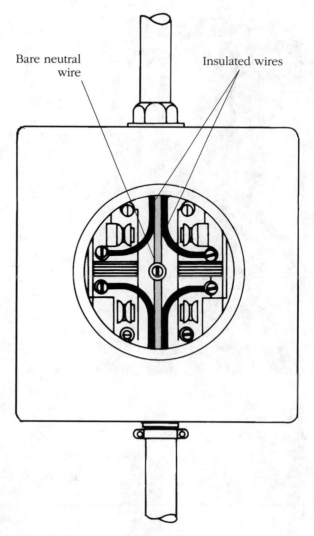

3-23 Cable connection in meter base.

Ground wires must be clamped tightly to the ground rod. Use brass or malleable iron clamps. A clamp such as the one shown in FIG. 3-36 works well with ground rods, while the clamp in FIG. 3-37 is often used on water pipes. Where rigid conduit enters the service entrance panel, fit the end with a grounding bushing (FIG. 3-38). In FIG. 3-39, a ground jumper wire is connected and run over to the neutral busbar.

TYPES OF BOXES

Outlet boxes used in the home wiring system come in two different materials. Rugged plastic boxes are called nonmetallic boxes; metal boxes are made of galvanized sheet steel (FIG. 3-40).

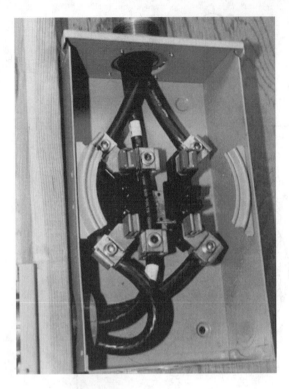

3-24 Meter base wired with unbroken neutral.

3-25 Use antioxidant compound when making connections with aluminum cables.

3-26 Cable connections should be neat and secure.

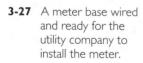

3-27 A meter base wired
and ready for the
utility company to
install the meter.

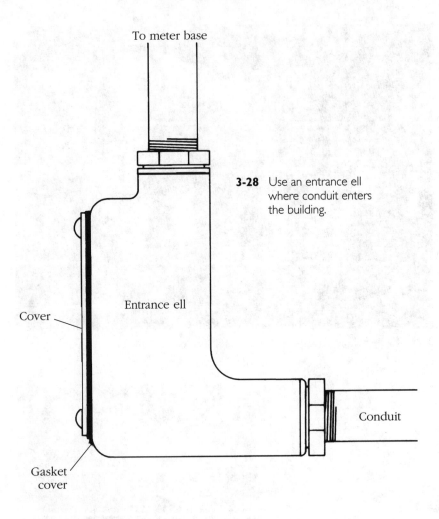

To meter base

3-28 Use an entrance ell where conduit enters the building.

Entrance ell

Cover

Conduit

Gasket cover

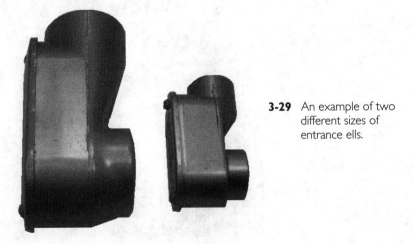

3-29 An example of two different sizes of entrance ells.

3-30 Cover gasket provides moisture protection.

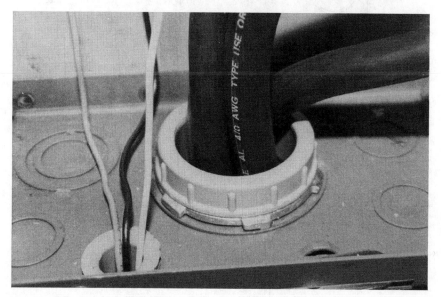

3-31 Cables protected by bushing.

Metal boxes

The metal box shown here (FIG. 3-41) is called a *switch box*; however, more duplex receptacles are installed in them than switches. The boxes have mounting holes to drive nails through into studs (FIG. 3-42). The boxes are equipped with mounting ears that adjust to make mounting easier (FIG. 3-43). The boxes also have marks on the sides corresponding to the thickness of the finished wall. Align these marks with the edge of the stud before you nail the box in place (FIG. 3-44). Extend the mounted box past the edge of the stud far enough so that the front edge of the box will be flush with the finished wall (FIG. 3-45).

Metal switch boxes may be ganged together, or they can be purchased already connected (FIG. 3-46). These boxes come with mounting brackets.

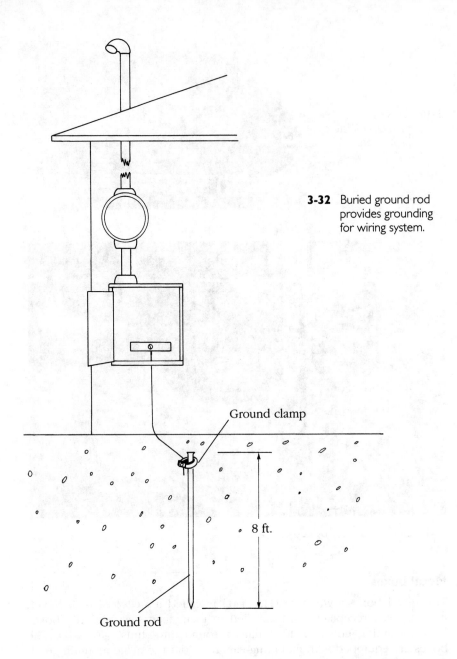

3-32 Buried ground rod provides grounding for wiring system.

Ground clamp

8 ft.

Ground rod

Some metal switch boxes have side-bracket supports. As the two screws are tightened (FIG. 3-47), the bracket folds and presses against the back of the wall.

Metal outlet boxes are available in octagonal or square shapes (FIG. 3-48). The octagon-shaped box is commonly used for ceiling light fixtures. Some have mounting brackets equipped with barbs. These help hold the box in place until you can drive the nails.

3-33 Ground wire connected to the neutral busbar in the service entrance panel.

3-34 Ground wire connected to the ground rod. Fill dirt will later completely bury rod.

If the ceiling outlet is to be mounted between joists, use a solid or adjustable bar hanger. In FIG. 3-49, the center knockout is removed from the back of the box, then the box is fastened to the hanger by the hanger's mounting bracket. Do not tighten the screw until the hanger is adjusted to the distance between the joists, then nail the hanger in place (FIG. 3-50).

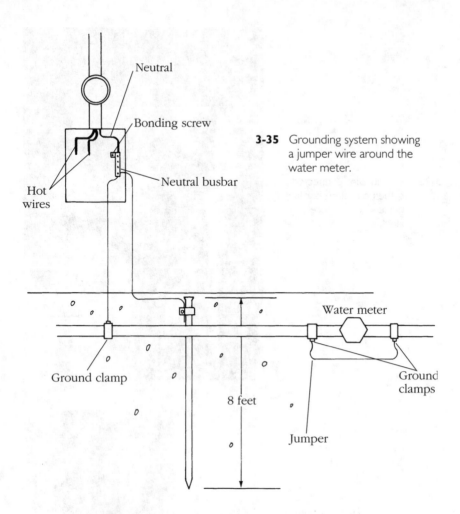

Neutral

Bonding screw

Hot
wires

Neutral busbar

3-35 Grounding system showing a jumper wire around the water meter.

Water meter

Ground clamp

Ground clamps

8 feet

Jumper

3-36 Ground clamp used on the ground rod.

3-37 Ground clamp used on the water pipe.

3-38 Rigid conduit with grounding bushing.

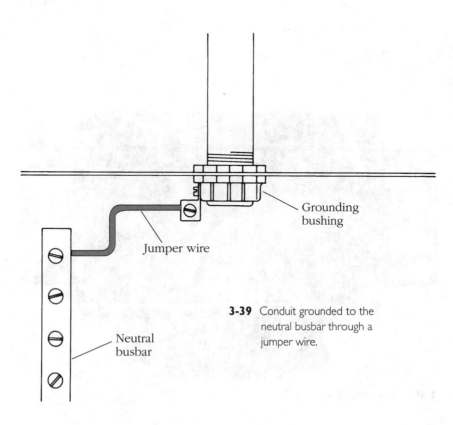

Grounding bushing

Jumper wire

Neutral busbar

3-39 Conduit grounded to the neutral busbar through a jumper wire.

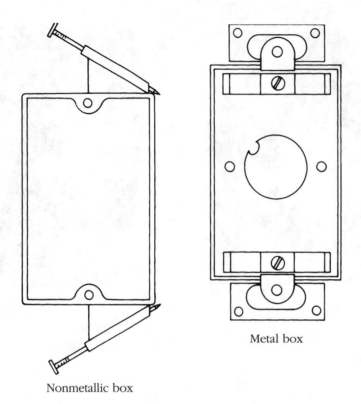

Metal box

Nonmetallic box

3-40 Nonmetallic and metal switch boxes.

3-41 Metal switch box.

3-42 The holes in back of the box are for mounting.

3-43 The slot in the ears allows for adjustment.

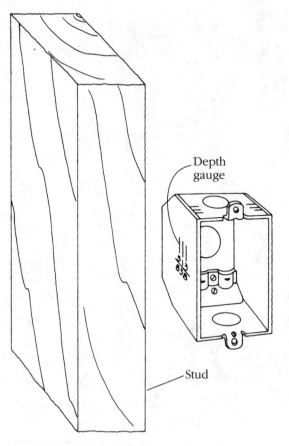

Depth
gauge

Stud

3-44 Align the selected mark on the edge of the stud.

3-45 Extend the front edge of the box past the stud to allow for the wall's thickness.

Nonmetallic boxes

Nonmetallic switch (plastic) boxes are widely used because they are inexpensive. They are used with nonmetallic sheathed cable, and they come equipped with the nails for quick mounting to studs (FIG. 3-51). They also

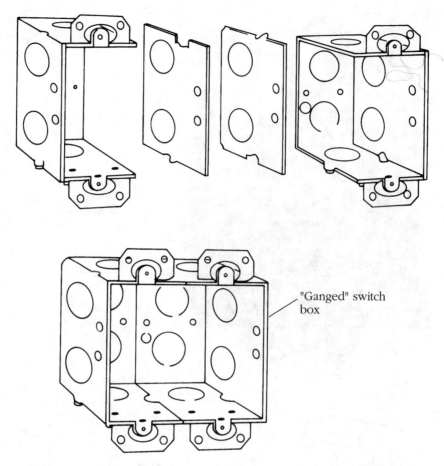

3-46 Metal boxes can be ganged together on the job.

"Ganged" switch box

have marks that can be used as guides to make sure the front of the box is flush with the finished wall. Nonmetallic boxes are also available in two-gang sizes (FIG. 3-52) or round-shaped sizes (FIG. 3-53). The round ones are nailed to ceiling joists for installing light fixtures.

Plastic boxes are durable, but they are breakable. When driving nails, be careful to hit the nails and not the box.

Inside plastic boxes you might notice a stamped number in cubic inches (FIG. 3-54). This refers to the inside dimensions of the box. For any size box there is a maximum number of wires allowed (TABLE 3-3). Keep in mind, though, that this is usable space, and you must make deductions for the cable clamps or straps, studs, and hickeys you would use in light fixtures. Reduce the allowable number of wires by one to make room for each of the extra items. Extra-deep boxes are available, if necessary. The point is not to crowd too many wires into too small a box.

3-47 The side bracket folds when the screw is tightened.

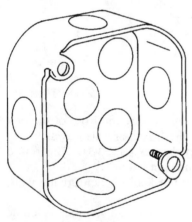

3-48 Larger outlet boxes are available in octagonal (left) or square (right) shapes.

3-49 An adjustable bar hanger.

Adjustable hanger

Hanger with box

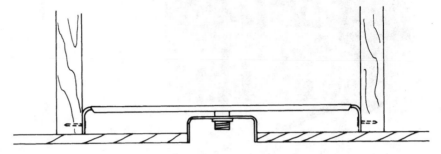

Hanger with box mounted
between joists

3-50 Use an adjustable hanger for mounting a box between joists.

3-51 Nonmetallic box ready
for nailing.

3-52 A nonmetallic two-ganged box.

3-53 A round nonmetallic box.

3-54 A nonmetallic box with cubic inches stamped in back.

Table 3-3

Type of box	Box size in inches		Maximum number of wires			
			No. 14	No. 12	No. 10	No. 8
Outlet box:	4 × 1¼	Round	6	5	5	4
	4 × 1½	or	7	6	6	5
	4 × 2⅛	Octagonal	10	9	8	7
	4 × 1¼	Square	9	8	7	6
	4 × 1½	Square	10	9	8	7
	4 × 2⅛	Square	15	13	12	10
	4¹¹⁄₁₆ × 1¼	Square	12	11	10	8
	4¹¹⁄₁₆ × 1½	Square	14	13	11	9
	4¹¹⁄₁₆ × 2⅛	Square	21	18	16	14
Switch box:	3 × 2 × 1½		3	3	3	2
	3 × 2 × 2		5	4	4	3
	3 × 2 × 2¼		5	4	4	3
	3 × 2 × 2½		6	5	5	4
	3 × 2 × 2¾		7	6	5	4
	3 × 2 × 3½		9	8	7	6
Handy box:	4 × 2⅛ × 1½		5	4	4	3
	4 × 2⅛ × 1⅞		6	5	5	4
	4 × 2⅛ × 2⅛		7	6	5	4

Mounting switch & outlet boxes

It is common practice to mount receptacle boxes back-to-back when the outlets serve two different rooms (FIG. 3-55). When mounting boxes, maintain the proper distance from the floor, secure the boxes, and don't forget to allow for the thickness of the wall.

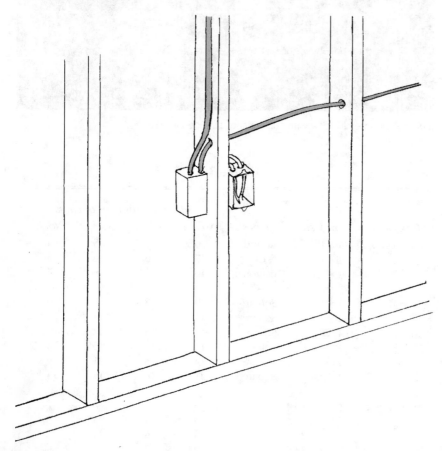

3-55 Receptacle boxes mounted back-to-back.

Chapter 4

Drilling holes & pulling wire

Most home wiring installations are made with nonmetallic sheathed cable. Some sort of protected path must be provided from the service entrance panel to the outlets. The cable must be protected from nails or other types of damage it might encounter after it has been installed. You can provide such protection by running the cable through drilled holes or notches in the studs and joists. In attics, you can run the cable between guard strips.

An electric drill is normally used to drill the holes. Flat wood bits are available in sizes ¼ inch and larger. A ⅜-inch bit is large enough for most of the holes. You can drill holes through top plates to feed cable down to switch boxes and receptacle outlets. Keep the holes in line to make it easier to pull the cable through. Center the hole in the stud (FIG. 4-1).

STEEL PLATES

The Code requires a minimum of 1¼ inches between the outside edge of the hole and the edge of the stud (FIG. 4-2). A ⅜-inch hole in a 2-x-4-inch (1¾-x-3¾-inch) stud will have to be positioned dead-center to leave enough on each side. If the hole is large or close to the edge, install a ¹⁄₁₆-inch steel plate (FIG. 4-3). These plates provide protection against nails when wallboard is being installed. Holes drilled in the center do not weaken the studs. Notches can be used, provided that they are not deep enough to weaken the structure (FIG. 4-4). Also cover these notches with a ¹⁄₁₆-inch steel plate (FIG. 4-5). Again, the purpose is to make sure the cable is protected against damage from nails.

Where the cable will interfere with wall insulation, cut notches in the bottom of the studs (FIG. 4-6). The problem with notches is that they must

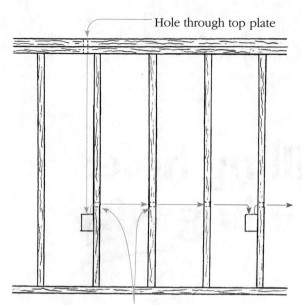

Hole through top plate

Holes through studs to feed receptacles

4-1 Keep holes in line to make cable easier to pull.

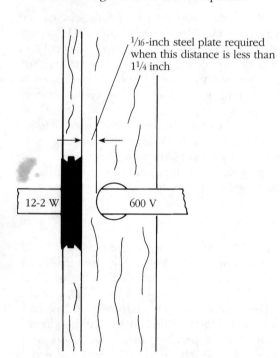

$1/16$-inch steel plate required when this distance is less than $1\frac{1}{4}$ inch

12-2 W   600 V

4-2 If the outside edge of a hole is less than $1\frac{1}{4}$ inches from edge of stud, it must be protected.

be cut before the stud is installed, which is usually not practical, unless you are also the one doing the framing. The other option is to drill the holes close to the bottom of the stud and run the cable there (FIG. 4-7). In either case, the insulation will be easier to install and be more efficient.

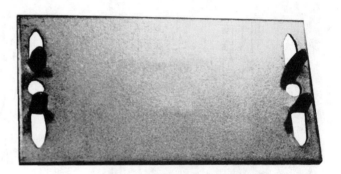

4-3 A ⅟₁₆-inch steel plate is available from electrical suppliers.

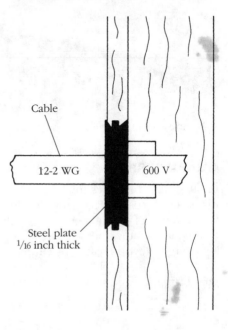

4-4 A sawed notch is easily removed with a wood chisel.

4-5 Cover the notch with a steel plate.

Cable

12-2 WG   600 V

Steel plate
¹/₁₆ inch thick

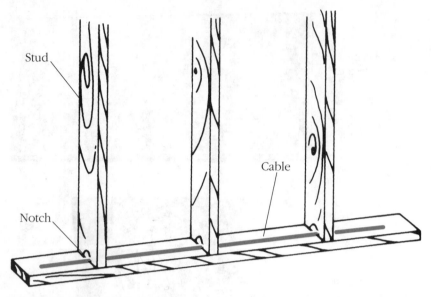

4-6 Notches cut in the bottom of the stud.

4-7 Holes drilled in the bottom of the stud.

GUARD STRIPS

Cable installed in attics might run across the top of the joists. If the attic is not easily accessible and not floored, the cable need only be protected within 6 feet of the attic entrance. If there are stairs or a permanent ladder leading to the attic, the cable must be protected.

You can protect the cable by running it through drilled holes in the joists (FIG. 4-8). If the cable runs across the top of the joists, you can install guard strips at least as high as the cable is thick (FIG. 4-9).

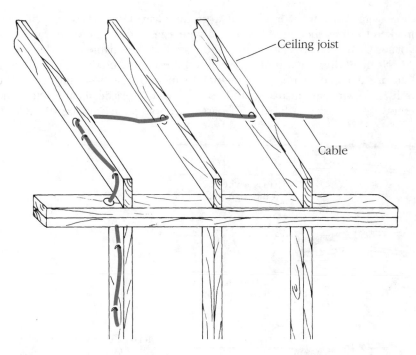

4-8 Run cable through drilled holes in ceiling joists.

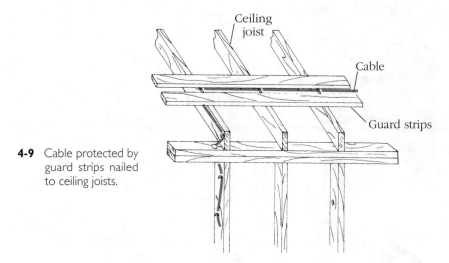

4-9 Cable protected by guard strips nailed to ceiling joists.

TYPES & SIZES OF CABLE

Cable made up of copper wires is much preferred over aluminum because the former is a better conductor and creates fewer problems later. In large sizes, however,—such as in service entrances—aluminum is often used because it is less expensive. Cable used in house wiring is usually two- or three-wire cable with or without a ground wire (FIG. 4-10). The cable is marked to show the type of wire, the wire size, and the maximum working voltage (FIG. 4-11). It will also show the manufacturer's name or trademark.

Nonmetallic sheathed cable is commonly used in house wiring because it makes a neat, clean—not to mention easy—installation. The individual wires are insulated and covered with a plastic sheath. Often the empty space inside is filled with paper wrapping, such as the #12-2 with

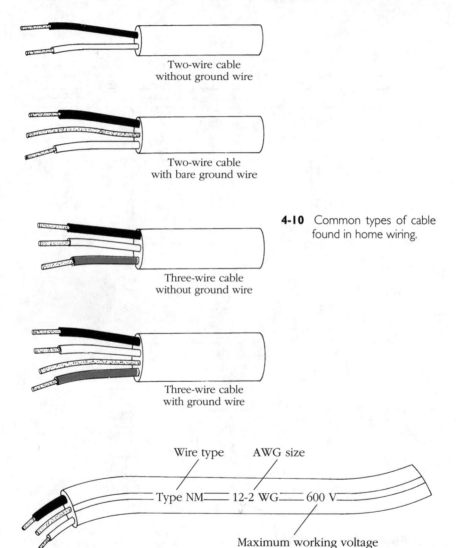

Two-wire cable
without ground wire

Two-wire cable
with bare ground wire

4-10 Common types of cable found in home wiring.

Three-wire cable
without ground wire

Three-wire cable
with ground wire

Wire type   AWG size

Type NM   12-2 WG   600 V

Maximum working voltage

4-11 Nonmetallic sheathed cable with identifying marking.

4-12 Nonmetallic cable No. 12-2 with ground, cable in plastic sheath.

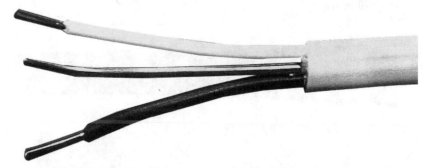

4-13 Nonmetallic cable No. 12-2 with ground, with sheath and paper wrapping cut away.

4-14 No. 12-3 with ground, cable showing paper wrapping.

ground (FIG. 4-12). The wrapping must be neatly cut away when making connections (FIG. 4-13).

The same size cable is available with three wires and a ground (FIG. 4-14), and this cable also contains insulation that must be removed. The Code calls this type of cable *NM*. This type of cable is only to be used in dry locations.

Type *NMC* cable does not have the paper wrapping. The individual wires are embedded in a solid sheath. NMC cable can be used in damp or corrosive locations indoors or out, but it cannot be exposed to the weather or buried.

For cable to be buried (other than service entrance), it must be type *UF*. As you can see from FIG. 4-15, this type also has the wires embedded in a solid sheath. It can be used for direct burial underground as well as where NM and NMC cable is used. It must, however, be protected at the starting point by fuses or a circuit breaker.

Number 8 or larger sizes of wire is often *stranded* (FIG. 4-16). This means a number of smaller wires are grouped together to form a larger, more flexible wire.

4-15 No. 12-2 with ground type UF cable.

4-16 No. 8-3 with round cable.

PULLING THE WIRE

Nonmetallic sheathed cable is usually available in 250-foot-long rolls of cable, which comes packaged in a carton with a perforated circle marked on top. To remove the cable, cut out the circle and feed the cable from the inside of the roll, rotating the end of the cable against the direction of the coil (FIG. 4-17). In this way, you can remove the cable without having it bind against itself.

You can start the cable either at the service entrance panel or the outlet to be wired. The usual choice is to start from the service entrance panel. Place the carton of wire on the floor beneath the service entrance panel. Pull the cable from the carton through the holes or notches to the first outlet for that circuit. You will probably find it very helpful to have a friend feed the cable while you pull. Feed the cable into the first outlet, allowing 6 or 8 inches to make connections.

Metal boxes have pry-out holes (FIG. 4-18), or you might need to use one of the larger knockouts (FIG. 4-19). Secure the cable to each metal box

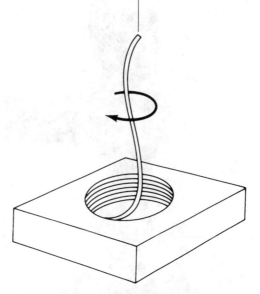

4-17 Cable being removed from carton.

4-18 Pry out the opening in the box with a screwdriver.

4-19 Remove the knockout with a sharp rap from hammer.

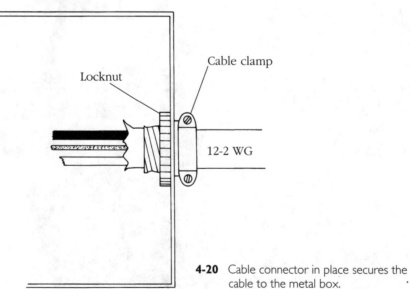

Locknut

Cable clamp

12-2 WG

4-20 Cable connector in place secures the cable to the metal box.

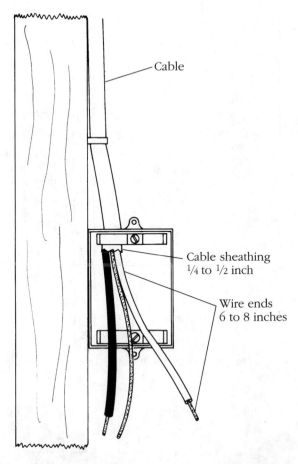

Cable

Cable sheathing
1/4 to 1/2 inch

Wire ends
6 to 8 inches

4-21   About 6 to 8 inches of wire should extend from the box to make connections.

with a clamp or a connector. If you use a connector, slide it over the cable end about 8 inches and tighten the screws. If the cable sheathing has already been removed, the cable must extend past the clamp or connector 1/4 to 1/2 inch.

Now insert the cable through the knockout and install the locknut (FIG. 4-20). After the cable is installed in the box, there should be 6 to 8 inches left to make the connections (FIG. 4-21).

The Code does not require the cable to be fastened to nonmetallic, single-gang switch boxes, but it must be fastened to all other boxes.

STAPLING THE CABLE

Strap or staple the cable within 8 inches of nonmetallic boxes and within 12 inches of metal boxes (FIG. 4-22). Further fasten the cable to support members every 4½ feet. Staples should be snug to the cable; do not drive

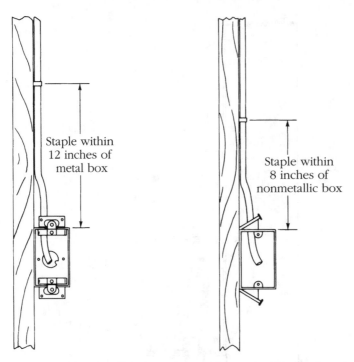

4-22 Anchor the cable within 12 inches of a metal box and within 8 inches of a non-metallic box.

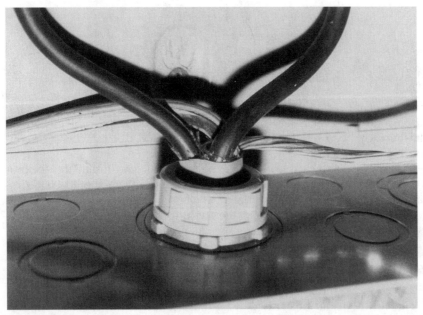

4-23 Cables entering the service entrance panel are protected by bushings.

them hard enough to crush the insulation. After the cable has been installed in the box and the sheathing has been cut away, fold the wires back in the box.

Back at the service entrance panel, allow about 3 feet of cable to extend inside the cabinet to make neat, uncluttered connections, then cut the cable. Protect the cables that enter the panel with bushings (FIG. 4-23).

Next, move the carton from the service entrance panel to the first outlet (that was just roughed in) and pull the cable to the second outlet. This process is simply repeated until the last outlet of the first circuit is completed. At that point, mark or tag the cable end in the service entrance panel to indicate the outlets and in what rooms the circuit serves; for example: LIGHTS, KITCHEN, & DINING ROOM. Repeat the steps until all of the circuits have been roughed up. The process of wiring a house will vary depending on the convenience of each situation and the number of people doing the job.

Chapter **5**

Making connections at the outlets

Before any connections can be made, you must remove the outer covering of the cable. To do this, cut down the center of the cable. A knife will work, but there is a risk of cutting the insulation on the wires. A cable ripper works better and faster (FIG. 5-1).

Next, cut away the sheath and paper filler with a knife or side cutter, allowing ¼ inch to extend inside the box (FIG. 5-2). Depending on the connection, you'll want to remove ½ to ¾ inch of insulation. If you use a wire stripper with notched jaws, simply match the notch with the wire size, place the wire in the notch, and squeeze the handle. A gentle pull outward will strip the insulation from the wire (FIG. 5-3). If you use a knife, make the cut at an angle to prevent nicking the wire. Nicked wires are weaker and cannot carry as much current (FIG. 5-4).

CONNECTING WIRES

When two or more wires are joined together, a solderless connector is normally used. These connectors are available in sizes to handle most wires. The connection must be good, allowing it to conduct electricity as well as an uncut wire.

Some connectors are crimped on, but the most common one—usually called a *wire nut*—is screwed on. The inside of the wire nut has threads (FIG. 5-5).

After you have removed the insulation, hold the ends of the wires together and simply screw the wire nut on (FIG. 5-6). The wires will twist together as the wire nut is screwed over the ends.

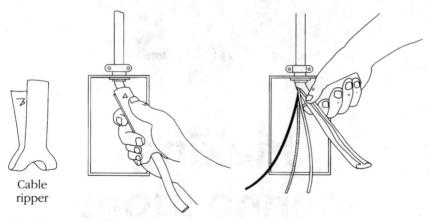

Cable ripper

5-1 Use an inexpensive cable ripper to strip the sheath from nonmetallic sheathed cable.

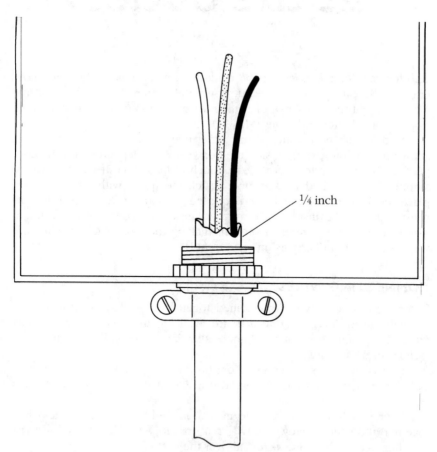

¼ inch

5-2 Always allow ¼ inch of sheath to show inside the box.

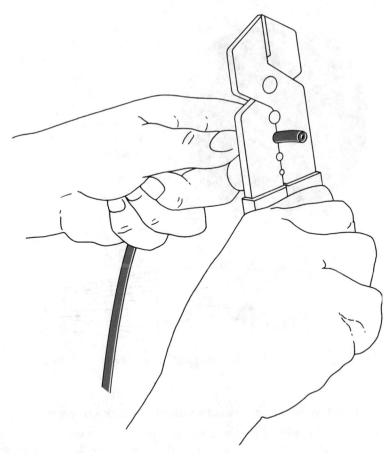

5-3 Wire strippers will quickly remove the insulation from wires when making connections.

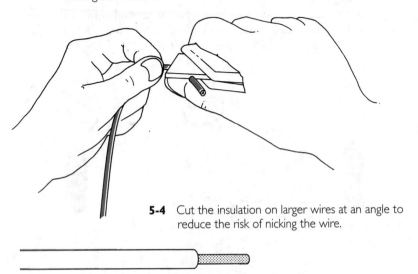

5-4 Cut the insulation on larger wires at an angle to reduce the risk of nicking the wire.

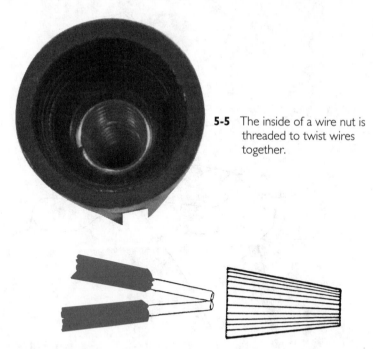

5-5 The inside of a wire nut is threaded to twist wires together.

5-6 It is not necessary to twist wires together. The wire nut will do that when it is screwed on.

Connecting wires to screw terminals & back-wiring devices

When you connect a wire to a screw terminal, remove only enough insulation to allow a loop to form around the screw (FIG. 5-7). Form the loop clockwise so that when the screw is tightened, the screw will not push the wire out of place.

You also can wire most switches and receptacles from the back. Simply push the bare end into the proper hole (FIG. 5-8).

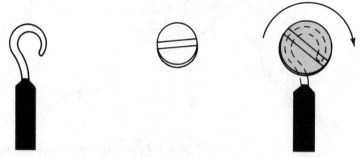

5-7 Install a wire loop in the same direction you turn the screw to tighten.

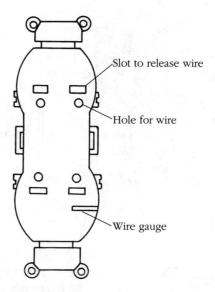

Slot to release wire

Hole for wire

Wire gauge

5-8 Wires are clamped when they are inserted in the hole. To release the clamp, use the slot.

Connecting ground wires to boxes & receptacles

To connect the ground wire to a grounding screw in a metal box, a *pig-tail* is normally used (FIG. 5-9). When a screw is not available, you can use a grounding clip (FIG. 5-10).

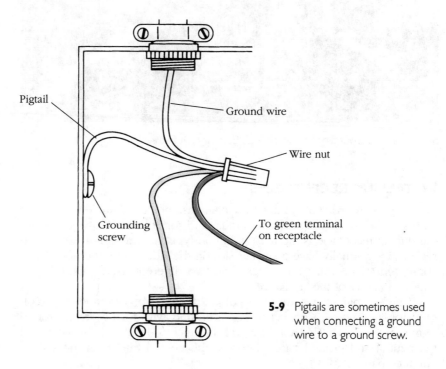

Pigtail

Ground wire

Wire nut

Grounding screw

To green terminal on receptacle

5-9 Pigtails are sometimes used when connecting a ground wire to a ground screw.

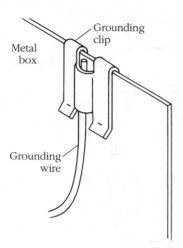

Grounding
clip

Metal
box

Grounding
wire

5-10 A grounding clip can be used instead of a screw.

The grounding wire does not have to be fastened to single-gang plastic boxes. Round plastic boxes used for light fixtures have a metal grounding bar (FIG. 5-11). If the light fixture is in the middle of the run, use a grounding pigtail; if the fixture is at the end of a run, simply connect the ground wire to the metal bar.

5-11 Use the grounding bar on a plastic box for light fixtures.

INSTALLING RECEPTACLES

Receptacles are available in a variety of styles, including single receptacles with switches or pilot lights. The standard 15- or 20-amp, 120-volt grounding duplex receptacle is the most common type, however (FIG. 5-12). Such receptacles normally have a smaller slot on the side where the hot wire is connected (FIG. 5-13). Some plugs have two different-sized pins to maintain the polarity of the appliance.

The terminal screws are brass-colored for the hot wire and silver-colored for the neutral. Connect the bare ground wire to the green ground terminal (FIG. 5-14). The hot wire will then go to the circuit breaker, and the neutral and ground wire will go to the neutral busbar in the service entrance panel (FIG. 5-15).

5-12 A standard duplex receptacle.

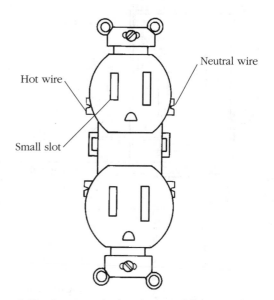

Hot wire

Neutral wire

Small slot

5-13 A receptacle showing two different sizes of slots.

5-14 A bare ground wire connected to ground terminal of a receptacle.

If you prefer back wiring, use the strip gauge on the back of the receptacle for removing the insulation. Then simply insert the end in the proper hole (FIG. 5-16). To remove the wire, press the end of a small screwdriver into the slot next to the hole (FIG. 5-17) to release the wire.

When the receptacle is at the end of the run, you have the only two insulated wires and the ground wire to connect (FIG. 5-18). When the receptacle is in the middle of the run, you can connect the four insulated wires to the four terminal screws, or into the holes for back wiring, or into a pigtail arrangement (FIG. 5-19).

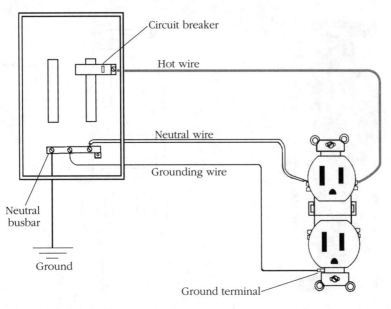

5-15 Wire connections from a receptacle to the service entrance panel.

Circuit breaker

Hot wire

Neutral wire

Grounding wire

Neutral busbar

Ground

Ground terminal

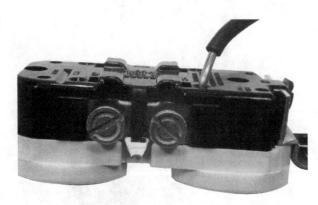

5-16 Back view of a receptacle shows the terminal screws, the holes used for back wiring, and the strip gauge.

5-17 Use a small screwdriver to remove the wire.

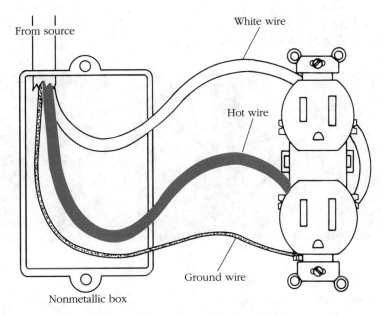

From source

White wire

Hot wire

Ground wire

Nonmetallic box

5-18 A receptacle installed at the end of a run.

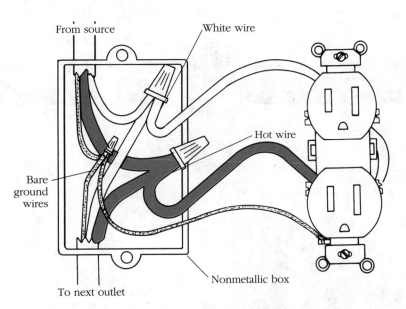

From source

White wire

Hot wire

Bare ground wires

To next outlet

Nonmetallic box

5-19 Pigtails used to wire a receptacle in the middle of a run.

INSTALLING LIGHT FIXTURES

Most light fixtures come equipped with factory wiring; however, some fixtures that mount directly to the box might have two terminal screws. For these fixtures, connect the hot wire to the brass-colored screw and the neutral wire to the silver-colored screw. Connect the bare ground wire to the grounding screw on the box (FIG. 5-20).

5-20 Wiring to a light fixture installed on a plastic box.

When using a metal box, remove one of the knockouts in the side. If the box is to be installed on a hanger, use the back knockout and hold the box in place with the bracket on the hanger (FIG. 5-21). After the bracket is adjusted for fit, nail it to the joist or stud (FIG. 5-22).

5-21 Use a mounting bracket to attach the box to the hanger.

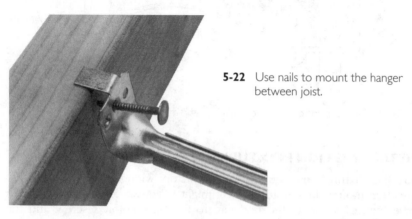

5-22 Use nails to mount the hanger between joist.

One type of fixture is attached to a strap mounted on the box (FIG. 5-23). After the wires are connected, mount the strap on the box, then install the threaded studs in the strap. Put the fixture in place and secure it with ornamental nuts.

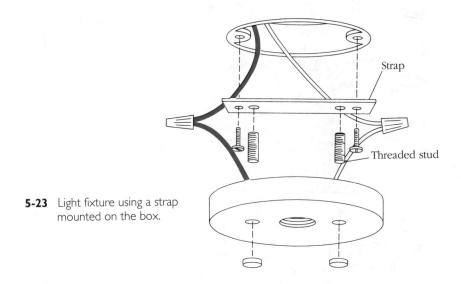

5-23 Light fixture using a strap mounted on the box.

Another type of fixture uses a threaded stem and a coupling, sometimes called a hickey. In FIG. 5-24, the fixture wires are pulled through the threaded stem and coupling, and the stem is screwed into the coupling. Then the wires are connected.

INSTALLING SWITCHES

Switches and receptacles come with the mounting screws in place. They are held there by small squares of paper or plastic (FIG. 5-25). Leave these squares on the screws (otherwise, the screws have a way of getting lost). Switches are stamped with the amount of current and voltage to be used and whether copper or aluminum wire can be used. They are often installed ganged together along side doors leading outside, in hallways, and above kitchen countertops, or they might be installed with a duplex receptacle over a workbench (FIG. 5-26).

Toggle switch

The most common switch found in the home is the *toggle switch* or the decorator-style rocker switch. It can be a single-pole, three-way, or four-way switch, depending on the number of locations desired to control the light. A toggle switch can be identified by the number of terminals it has.

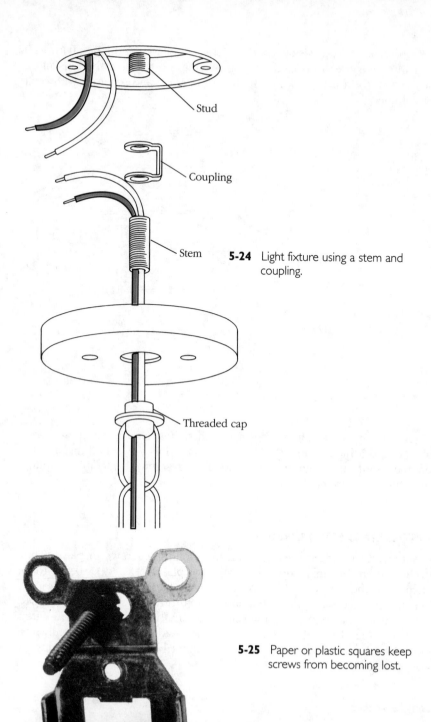

Stud

Coupling

Stem

5-24 Light fixture using a stem and coupling.

Threaded cap

5-25 Paper or plastic squares keep screws from becoming lost.

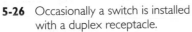

5-26 Occasionally a switch is installed with a duplex receptacle.

5-27 A common single-pole switch.

Most of the switches in the home are the single-pole type (FIG. 5-27). The color of the wires connected to the switch depends on whether the power goes to the light first or to the switch first. If power goes to the switch first, connect the two black wires (FIG. 5-28) or use a pigtail arrangement such as the ones shown (FIG. 5-29) and (FIG. 5-30). If the power goes to the light first, use a switch loop with the white wire as a hot wire (FIG. 5-31), though it is not necessary to identify the white wire as a hot wire. Other situations require the white wire to be marked with black tape or paint to show that it is being used as a black (or hot) wire.

Split receptacles

Often it is desirable to have a switch control one-half of a duplex receptacle. To do this, remove the tab between the two brass screws of the receptacle (FIG. 5-32), then connect the switch loop as shown in FIG. 5-33.

Three-way switches

For safety and convenience, you might want to be able to control lights from two locations on stairs, hallways, garages, and rooms with two entrances. To accomplish this, use two three-way switches (FIG. 5-34). **Note:** Despite their name, three-way switches alone will only control a light from two locations. The three-way description refers to the number of terminals on the switch.

Three-way switches have three screw terminals. Such switches are used only in pairs. These switches, along with having three screw terminals, also do not have ON/OFF markings (FIG. 5-35).

Most switches are purchased in sealed individual boxes. Their boxes indicate the type and size of switch inside. Some manufacturers include

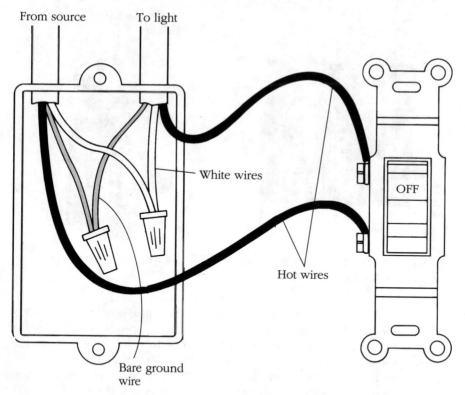

5-28 Two black wires connect to the switch.

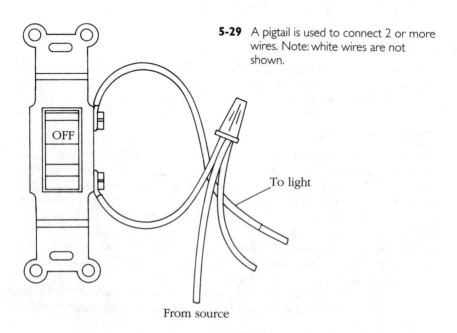

5-29 A pigtail is used to connect 2 or more wires. Note: white wires are not shown.

5-30 This switch is wired using black pigtail and white wire.

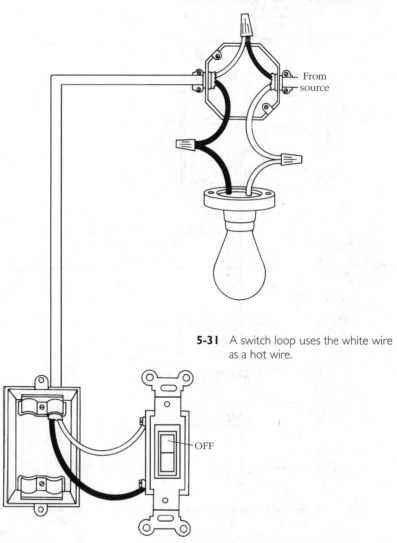

From source

5-31 A switch loop uses the white wire as a hot wire.

OFF

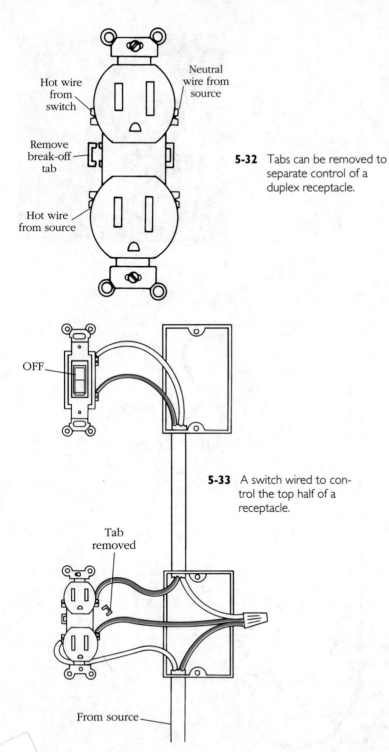

Hot wire from switch

Neutral wire from source

Remove break-off tab

Hot wire from source

5-32 Tabs can be removed to separate control of a duplex receptacle.

OFF

5-33 A switch wired to control the top half of a receptacle.

Tab removed

From source

Note: Ground wire is not shown

Making connections at the outlets

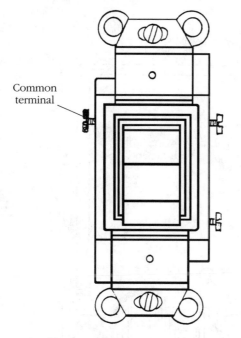

Common terminal

5-34 A three-way switch has one terminal darker than the others. This is the common terminal that must be used for control. The other terminals are used to connect the two switches.

5-35 Three-way switches have no ON/OFF markings.

wiring instructions. Wiring three-way switches is usually confusing at first because a third wire must be used between the switches. These switches usually can be back-wired.

Three-way switches are wired depending on the location of the light in relation to the switches and the power source. Figure 5-36 shows a wiring arrangement where the power goes to the light and then to the switches. Figure 5-37 illustrates a wiring procedure when the power is connected to a switch first and the light is at the end of the run. If the light is located between the two switches, it can be wired as shown in FIG. 5-38.

Four-way switches

To control a light from more than two locations, use a four-way switch (FIG. 5-39) with two three-way switches. Four-way switches have four screw terminals, and most can be back wired. They are used only with a pair of three-way switches. Use three-wire cable between all switches, and two-wire cable from the light to the first switch (FIG. 5-40). To control a light from more locations, install another four-way switch between the two three-way switches for each additional location.

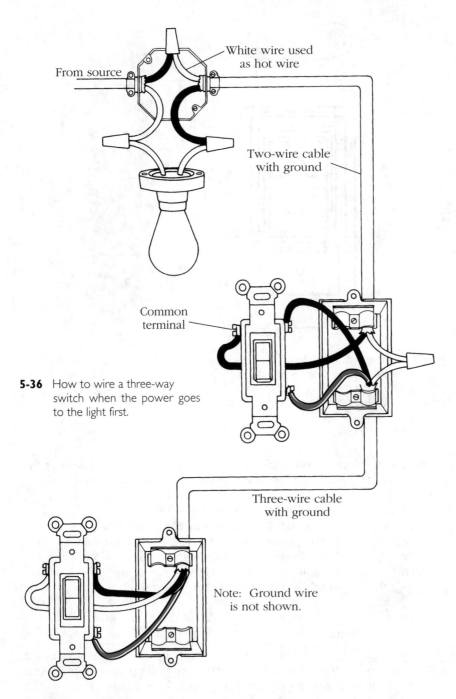

White wire used
as hot wire

From source

Two-wire cable
with ground

Common
terminal

5-36 How to wire a three-way
switch when the power goes
to the light first.

Three-wire cable
with ground

Note: Ground wire
is not shown.

Dimmer switches

If a varying level of light is desired, use a dimmer switch (FIG. 5-41). These
switches control the brightness as well as turning the light on and off.
Install them the same way as the regular single-pole switch (FIG. 5-42).

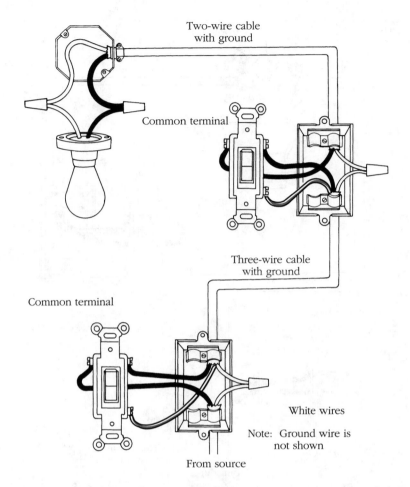

Two-wire cable
with ground

Common terminal

Three-wire cable
with ground

Common terminal

White wires

Note: Ground wire is
not shown

From source

5-37 How to wire a three-way switch when the power goes to the switch first.

INSTALLING GROUND FAULT CIRCUIT INTERRUPTERS

Circuit breakers, appliance grounding, and a properly grounded electrical system provide important protection against shock. Some locations are considered more hazardous, however, and require a more sensitive safety device. The ground-fault circuit interrupter (GFCI) was developed for this reason (FIG. 5-43). Some GFCIs are prewired at the factory (FIG. 5-44).

The Code requires that temporary outlets used on construction sites be protected by a GFCI along with some permanent outlets in the home. These include the receptacles installed in the bathroom, all outdoor receptacles, and those receptacles installed in a garage, except those used by an appliance such as a freezer or a garage door opener. A GFCI can be installed in a regular receptacle box. It has a TEST button and a RESET button, which should be used about once a month to ensure that the device is working.

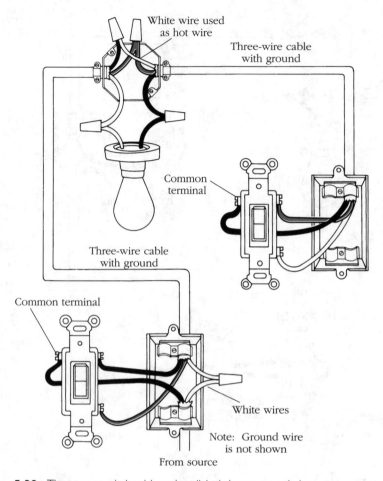

White wire used
as hot wire

Three-wire cable
with ground

Common
terminal

Three-wire cable
with ground

Common terminal

White wires

Note: Ground wire
is not shown

From source

5-38 Three-way switch wiring when light is between switches.

5-39 A typical four-way switch.

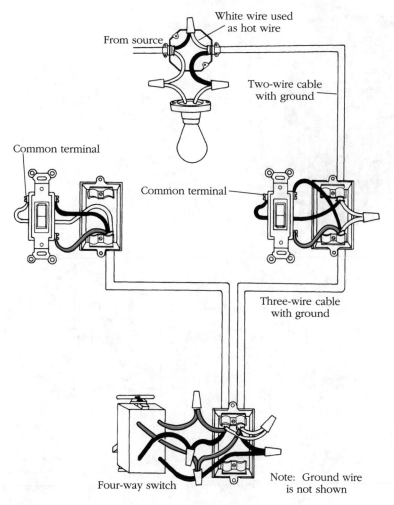

White wire used
as hot wire

From source

Two-wire cable
with ground

Common terminal

Common terminal

Three-wire cable
with ground

Four-way switch

Note: Ground wire
is not shown

5-40 Four-way switches are installed between two three-way switches to control a light from three or more locations.

Wire GFCIs the same as a standard receptacle, with the black wire going to the brass terminal and the white to the silver-colored terminal. Some GFCIs come with factory wiring but connect the same way—black wire to black and white to white. Most have provisions to connect to other regular receptacles (FIG. 5-45). This is an option that will provide GFCI protection to any receptacle installed downstream in the circuit.

INSTALLING CEILING FANS

Ceiling fans are wired just like light fixtures. Some models are light enough to be installed to an existing metal outlet box, but heavier fans

5-41 A typical dimmer switch with the control knob removed.

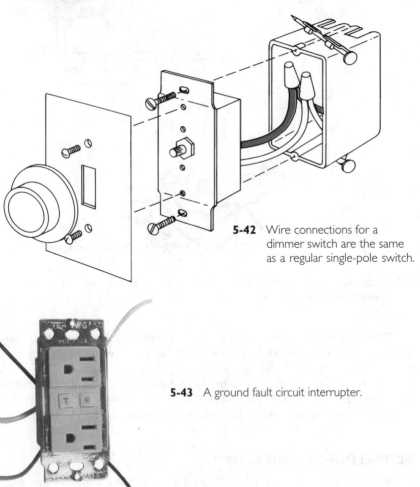

5-42 Wire connections for a dimmer switch are the same as a regular single-pole switch.

5-43 A ground fault circuit interrupter.

5-44 Back view of GFCI showing factory wiring.

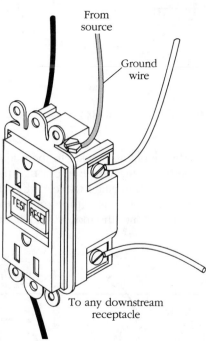

From source

Ground wire

TEST RESET

To any downstream receptacle

5-45 Wiring arrangement for a GFCI providing downstream protection If there are no other receptacles the bottom terminals are not used.

might need additional support. This usually means attaching the outlet box directly to a wooden joist or a brace between the joists. Remember not to place any strain on any of the wires.

Read the manufacturer's instructions carefully and review any diagrams thoroughly. Turn off the power to the circuit at the service entrance panel. Use the voltage tester to make sure the power is off, and then remove the existing fixture.

Mount the fan to a standard 4"-×-2⅛" metal octagon electrical box. The box must be firmly secured to a joist or reinforced to support the weight of the fan (at least 40 lbs.). Do not use a plastic outlet box. It probably would not support the fan (FIG. 5-46).

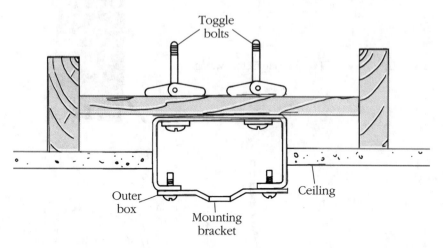

5-46 The box must be firmly secured to support the fan.

Securely fasten the mounting bracket to the outlet box using the screws and washers provided with the fan. Be sure that the mounting bracket is securely tightened against the ceiling to keep the fan from wobbling (FIG. 5-47). To mount the motor you must lift the motor and downrod to the outlet box and fit the hang ball into the mounting bracket (FIG. 5-48).

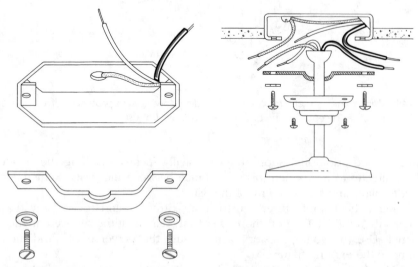

5-47 Fasten the mounting bracket securely to the box.

5-48 Fit the hang ball into the mounting bracket.

Use wire nuts to connect the house wires to the fan wires: black wire to black wire, white wire to white wire, and bare ground wire to green fan wire. Refer to the wiring diagram provided with the fan (FIG. 5-49). Fold the wires up into the outlet box and slide the canopy up the down-rod until it is flush with the ceiling. Tighten the set screws in the canopy. Attach one fan blade to the blade bracket using two blade mounting screws. Rotate the motor by hand until the blade is opposite you, then attach the second blade. Attach the remaining blades. Turn the power back on and try the fan. Normally, the directional switch should direct the air flow down for summer operation and up for winter.

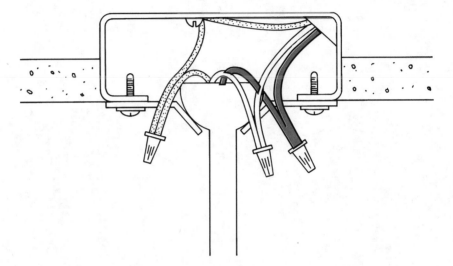

5-49 Connect the wires according to the wiring diagram provided with the fan.

Chapter **6**

Individual
appliance circuits

Most homes today have a number of built-in appliances that are permanently installed. They are usually referred to as *fixed appliances*. They include such items as ovens and countertop units, garbage disposals, and water heaters. Stationary appliances are those that are usually moved from one house to another, such as clothes dryers and self-contained ranges.

The Code does not require receptacles for built-in ovens or countertop units. They can be wired directly from the service entrance panel, providing the panel is protected by a disconnect such as a circuit breaker or a pull-out fuse block. For ease of servicing, however, a receptacle and plug is usually used.

Receptacles are rated by the current and voltage. Those rated at 125 volts may not be used for any voltage over 125. Receptacles rated at 250 volts may be used only for voltages above 125 but not more than 250. Figure 6-1 shows the type of receptacle used for 240-volt circuits. (The half-round openings seen at the bottom of the drawings are used for the equipment grounding prong and are never to be used for the neutral wire as with a 120/240-volt circuit.) Receptacles rated 125/250 volts are used only for appliances that operate on a combination 120/240 volts. These receptacles require a neutral wire running to the appliance.

MAKING 120/240-VOLT CONNECTIONS

Ranges operate at either 120 or 240 volts, depending on whether the control knob is turned to LOW, MEDIUM, or HIGH. The circuit for the range normally uses #6 three-wire cable and a 50-amp, 125/250-volt receptacle. These receptacles are designed to be surface-mounted (FIG. 6-2) or flush-mounted (FIG. 6-3). Some local codes require them to be flush-mounted

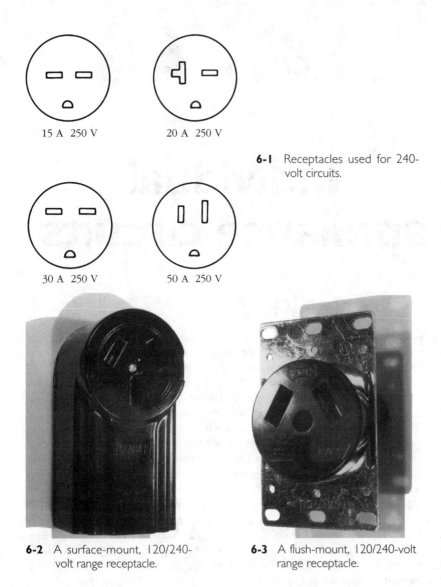

15 A 250 V

20 A 250 V

6-1 Receptacles used for 240-volt circuits.

30 A 250 V

50 A 250 V

6-2 A surface-mount, 120/240-volt range receptacle.

6-3 A flush-mount, 120/240-volt range receptacle.

and in metal boxes, so be sure to check with your area's regulations before installing your receptacle.

The 50-amp, 125/250-volt receptacles have three slots the same shape. On the surface-mounted type, a single screw holds the cover in place. Inside, there are three terminal screws and a cable clamp (FIG. 6-4). The back of the flush-mounted type has terminal screws and holes for the wires (FIG. 6-5).

A clothes dryer operates on 120/240 volts like the range. It is normally wired with #10, three-wire cable and a 30-amp, 125/250-volt receptacle. The receptacles are also available in surface-mounted or flush-mounted types. The 30-amp, 125/250-volt receptacle is the same as the 50-amp

6-4 Inside view of a 120/240-volt, surface-mount range receptacle.

6-5 Back view of a 120/240-volt, flush-mount receptacle.

range receptacle except the bottom slot is shaped like an inverted L (FIG. 6-6). The inside is also the same except for the L-shaped terminal.

The wire connections and procedures are the same for ranges and clothes dryers. The only difference is that the range will probably use #6,

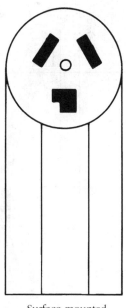

6-6 Dryer receptacle with L-shaped slot for the neutral prong.

Surface-mounted
receptacle

three-wire cable and a 50-amp receptacle, while the dryer will use a #10, three-wire cable and 30-amp receptacle.

To wire the receptacle, connect the white wire to the terminal marked W. The Code allows the white (neutral) wire in a three-wire cable to be used as a ground wire only for ranges and clothes dryers. The other two wires are the hot wires. They are usually red and black. Connect them to either of the other two terminals marked X and Y (FIG. 6-7).

The cable going to the appliance will have a plug (FIG. 6-8) to fit the proper receptacle. Some plugs come with an extra prong for the neutral terminal, which allows the plug to be used for either ranges or dryers. To wire this plug, loosen the four screws and separate the two halves (FIGS. 6-9, 6-10). Then wire the plug, making sure the neutral is connected to the center prong matching the receptacle.

MAKING 240-VOLT CONNECTIONS

Most water heaters operate on 240 volts. They are usually wired with #10-2 with ground cable to a 30-amp breaker. The cable might run directly to the water heater, or it might require a junction box (FIG. 6-11), depending on the local code.

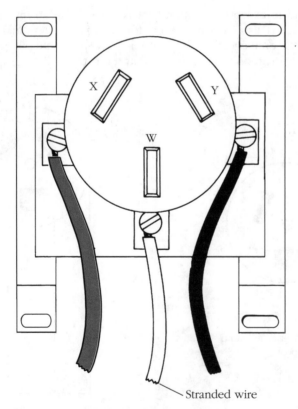

Stranded wire

6-7 Wire connections for 120/240-volt circuit.

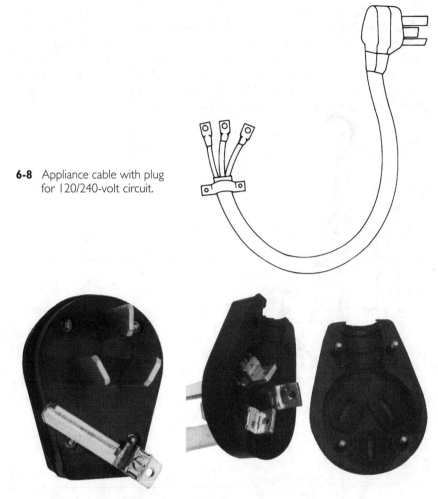

6-8 Appliance cable with plug for 120/240-volt circuit.

6-9, 6-10 The range plug can be modified to use on a clothes dryer.

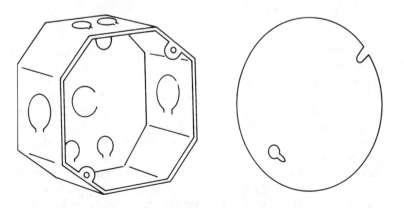

6-11 Junction box used to make connections to water heater.

Locate the junction box 12 or more inches above the water heater and wire it as shown in FIG. 6-12. In FIG. 6-13, there is only one black wire, one white wire, and one bare ground wire. The white wire will be used as a hot wire, so mark it with black tape to identify it. At the water heater, make the connections as shown in FIG. 6-13 and, again, mark the white wire to identify it as being used as a hot wire. Be sure to connect the bare ground wire to the metal case of the heater.

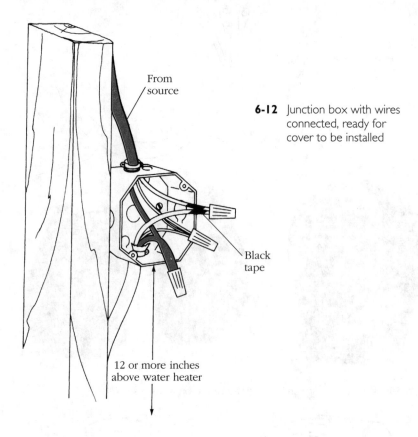

From source

6-12 Junction box with wires connected, ready for cover to be installed

Black tape

12 or more inches above water heater

When you buy service entrance panels, they are usually empty except for the main circuit breaker. The one shown in FIG. 6-14 has a 200-amp main breaker and has part of the circuit breakers installed. If you place the service entrance panel some distance from the meter, locate the main breaker next to the meter and connect the service entrance wires directly to the hot busbars (FIG. 6-15). Depending on how the cabinet was assembled, you can feed the entrance wires to the main breaker from the top or the bottom (FIG. 6-16).

Somewhere inside the cabinet is at least one neutral busbar (FIG. 6-17). Connect the white neutral and the bare grounding wires (FIG. 6-18) as well as the neutral conductor from the service entrance wires to the bar. Feed cables entering the service entrance panel through cable connectors (FIG.

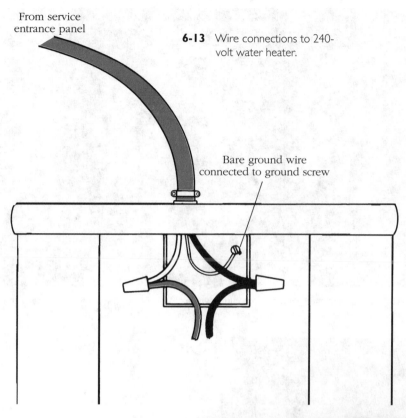

From service
entrance panel

6-13 Wire connections to 240-
volt water heater.

Bare ground wire
connected to ground screw

6-14 Service entrance panel with
200-amp main breaker and
assorted breakers.

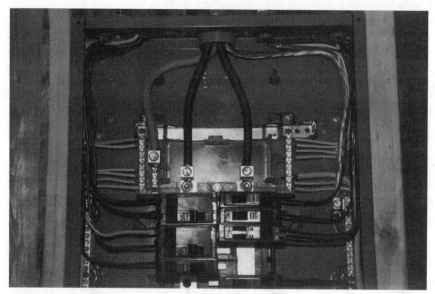

6-15 Service entrance without a main breaker. When the panel is located away from the meter, install the main breaker next to the meter.

6-16 A service entrance panel with the main breaker installed at the bottom of the cabinet.

6-19) and hold them in place with locknuts. Once inside, keep the wires to the sides of the cabinet until reaching the point where they will be connected (FIG. 6-20). This not only makes a more professional-looking installation, but it also makes troubleshooting simpler if you do not have to wade through a scramble of wires. In addition, such installation provides better air circulation and reduces heat buildup.

6-17 Neutral busbar in the service entrance panel.

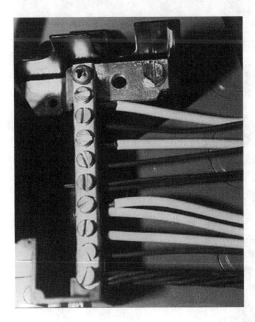

6-18 Both white neutral wires and bare ground wires are connected to the neutral busbar.

At this point, strip the sheath from the cables and extend the wires for each circuit inside the service entrance panel. Label each one as it is connected to the breakers, marking the location they serve on the chart inside the cabinet door.

MAKING SERVICE ENTRANCE PANEL CONNECTIONS

Most of the circuits will use single-pole breakers and a few will use double-pole breakers (FIG. 6-21). The breakers will be marked to identify their voltage rating along with the type and size of wire acceptable.

To connect the circuit, run the white wire and the bare ground wire to the neutral busbar. Remove about ½ inch of the insulation and insert the

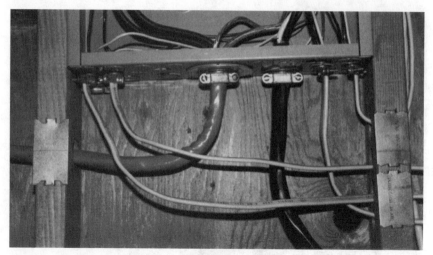

6-19 Insert cables entering the metal cabinet through cable connectors.

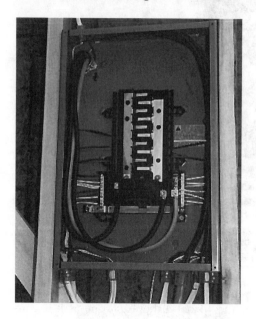

6-20 Run the incoming wires in an orderly fashion along the sides.

bare end in one of the terminal holes. Tighten the screw firmly. Connect the bare ground wire to another terminal. Next, measure and cut the black wire to length. Cut away about ½ inch of the insulation from the end. At one end of the breaker is the terminal screw, and beneath this screw is the hole where the wire is to go (FIG. 6-22). Insert the bare end of the wire in the hole, and then tighten the screw.

The procedure will vary, depending on the brand, but most circuit breakers plug into their terminals. Usually, you insert the breaker in a slot at one end, then push the cup end in and rotate it until it stops (FIG. 6-23). The finished wire connections should have a white wire and a bare

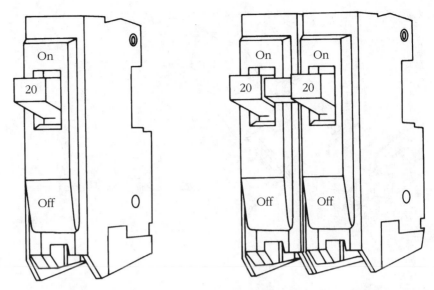

6-21 A single-pole breaker (left) used for 120-volt circuits and a double-pole breaker (right) used for 240-volt circuits.

6-22 Single-pole breaker showing the terminal screw. The hole for the wire is beneath the screw.

ground wire going to the neutral busbar and the black wire going to the circuit breaker (FIG. 6-24).

The double-pole circuit breaker is just like two single-pole breakers joined together (FIG. 6-25). The toggle part is connected so that both sides will trip together.

Most of the breakers will be single-pole with a black wire connected to supply the 120-volt circuits. The straight 240-volt circuits, such as the one to the water heater, will use one black wire and one white wire, with

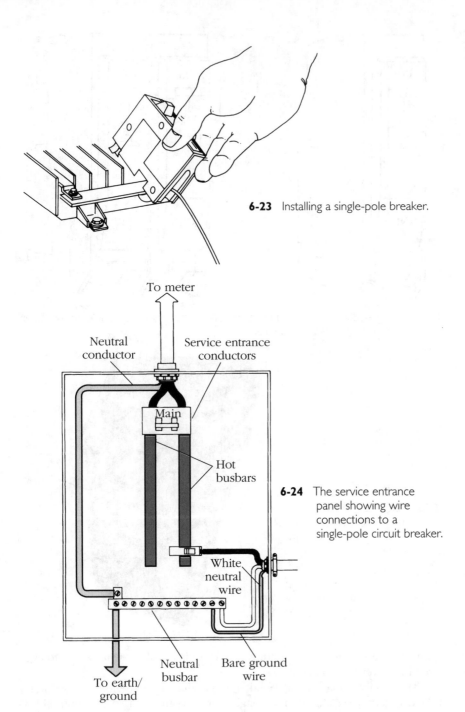

6-23 Installing a single-pole breaker.

To meter

Neutral
conductor

Service entrance
conductors

Main

Hot
busbars

6-24 The service entrance
panel showing wire
connections to a
single-pole circuit breaker.

White
neutral
wire

To earth/
ground

Neutral
busbar

Bare ground
wire

6-25 The terminal screws to a double-pole breaker.

the white wire marked black (FIG. 6-26). The 120/240-volt circuits for ranges and clothes dryers will be connected to a double-pole breaker with three-wire cable where the two hot wires will probably be red and black (FIG. 6-27). Figure 6-28 shows the wire connections for these three circuits.

INSTALLING CIRCUIT BREAKER GFCIs

To provide GFCI protection for an entire circuit, install a circuit breaker GFCI. GFCIs can be identified by the white pigtail wire and the PUSH-TO-TEST button (FIG. 6-29).

Install a GFCI by connecting the white neutral wire from the circuit to the terminal marked LOAD NEUTRAL and the black wire to the terminal marked LOAD POWER. These terminals might be located on the bottom of the breaker or on the end as shown in FIG. 6-30. Next, plug the breaker into its terminal and connect the curly, white wire to the neutral busbar along with the bare ground wire from the circuit. Figure 6-31 shows the wire connections.

The cover for the service entrance panel will have knockout blanks to accommodate each circuit breaker. Only remove the number of spaces necessary for the number of breakers installed. If any blanks are left, they may be removed later if another circuit is installed. When the cover is installed, the inside of the cabinet should be completely protected with only the tops of the circuit breakers exposed (FIG. 6-32).

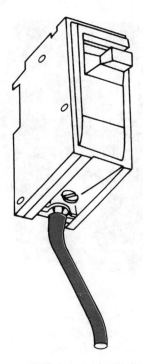

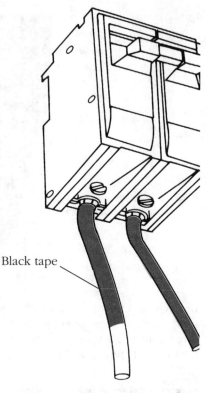

Black tape

Single-pole circuit breaker
with black wire connected
to a 120-volt circuit

Double-pole circuit brea
with wires connected to
240-volt circuit

6-26 A single-pole breaker using one black wire for 120 volts, and a double-pole breaker using one black wire and a white wire marked black for a straight 240 volts.

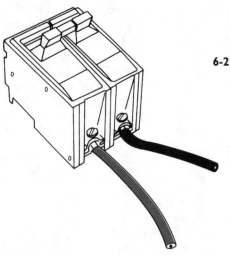

6-27 A double-pole breaker using a red and black wire for the 120/240-volt circuit.

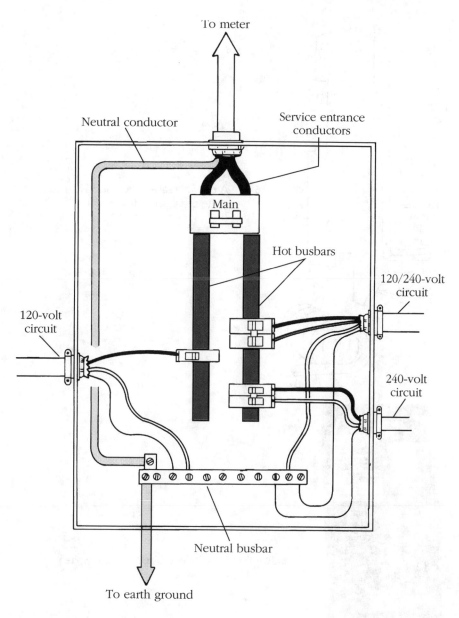

To meter

Neutral conductor

Service entrance conductors

Main

Hot busbars

120/240-volt circuit

120-volt circuit

240-volt circuit

Neutral busbar

To earth ground

6-28 A service entrance panel showing wire connections for a 120-volt, straight 240-volt, and a 120/240-volt circuit.

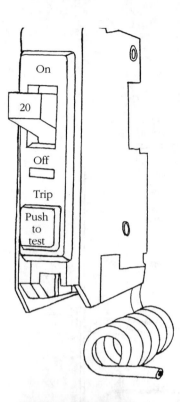

6-29 A circuit breaker GFCI will have a curly, white pigtail and a test button.

6-30 Terminals for one black and one white wire of a 120-volt circuit.

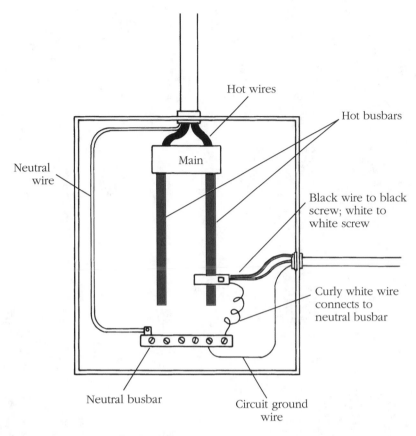

Hot wires

Hot busbars

Neutral
wire

Black wire to black
screw; white to
white screw

Main

Curly white wire
connects to
neutral busbar

Neutral busbar

Circuit ground
wire

6-31 Service entrance panel showing wire connections for a circuit breaker GFCI.

6-32 Service entrance panel
wired and with the cover
in place.

Chapter **7**

Installing special circuits

The installation of special circuits can provide conveniences throughout your home. Door chimes might announce a visitor at the door by playing part of a favorite tune. Thermostats can control central heating and cooling systems. Smoke detectors are now often required by Code, especially in new construction. And more and more homeowners are installing their own telephone systems. All of these conveniences require special circuits that the do-it-yourselfer can install.

DOOR CHIMES

Most building-supply stores have a display of a variety of door chimes. Regardless of how complicated the system, the wiring is pretty much the same. A transformer is used to step the house voltage down from 120 volts to a safer, lower voltage (FIG. 7-1). Older installations might have used only 6 to 10 volts (FIG. 7-2), while more modern chimes might require about 24 volts. If a back door needed a button, it was wired in as shown in FIG. 7-3.

Today's chimes are connected the same way, except they usually need a larger transformer (FIG. 7-4). If an older bell is replaced with a chime, and the chime doesn't perform as it should, the transformer will probably have to be upgraded as well.

The transformer is mounted in an outlet box, usually in a closet or utility room where it is visible. It should not be mounted in an attic where a problem might go unnoticed for some time. Check with your local building inspector.

Make the 120-volt connections inside the box—white wire to silver terminal and black wire to brass terminal. Use #18-2 or #20-2 insulated wire between the buttons, chime, and transformer. The chime consists of

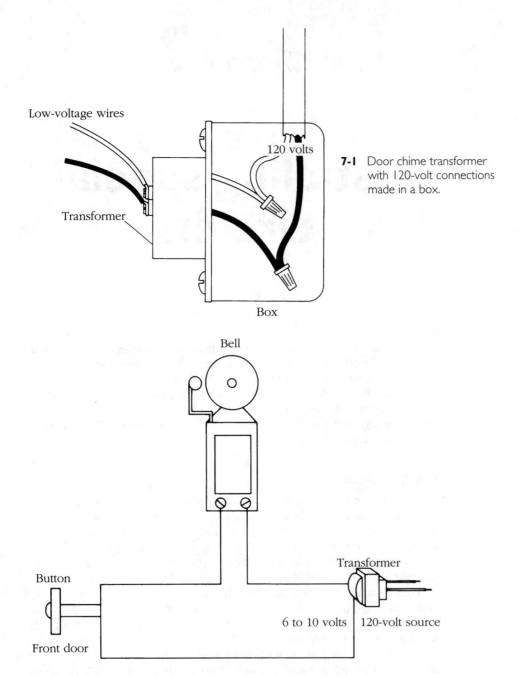

Low-voltage wires

120 volts

7-1 Door chime transformer with 120-volt connections made in a box.

Transformer

Box

Bell

Transformer

Button

6 to 10 volts | 120-volt source

Front door

7-2 Wiring diagram of a doorbell with one button.

two or more spring-loaded plungers (FIG. 7-5) that strike a bar or tube when the button is pushed. Most decorative covers just snap in place (FIG. 7-6). The push buttons will have two terminals where the wires can be connected to either screw (FIG. 7-7).

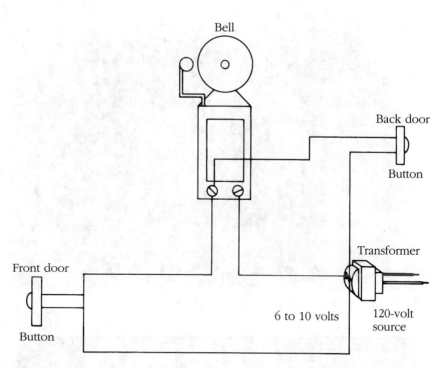

7-3 Wiring diagram of a doorbell with front and back door buttons.

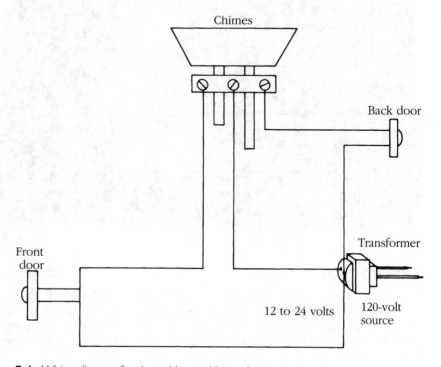

7-4 Wiring diagram for door chimes with two buttons.

7-5 Terminal connections in front of strikers for door chimes.

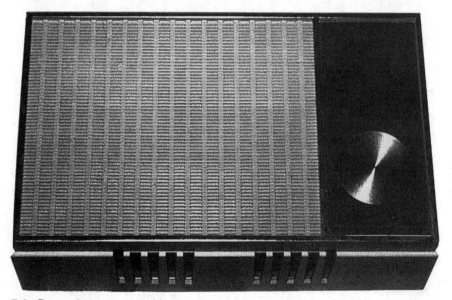

7-6 Decorative cover over chimes.

THERMOSTATS

Most homes today have central heating and cooling systems. These are controlled by a thermostat, which is really nothing more than a switch that responds to a change in temperature. The thermostat operates on the principle that two different metals expand in different amounts for the same temperature change.

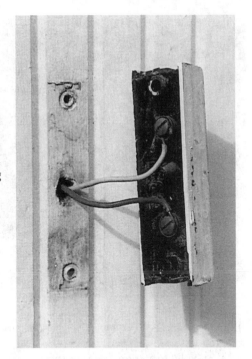

7-7 Door chime button showing two terminal screws.

Thermostats are made up of two strips of different metals bonded together, often in the shape of a coiled spring. As the temperature changes, the bonded strip slowly bends, creating the switching action that opens or closes the electrical contacts (FIG. 7-8).

Most heating and air-conditioning units are equipped with a low-voltage transformer (FIG. 7-9). The transformer lowers the control voltage to

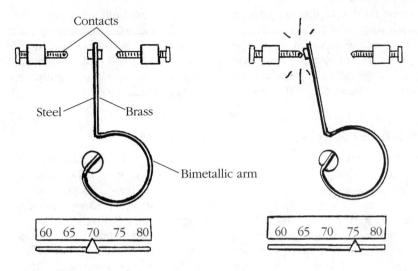

7-8 Bimetallic arm making contacts.

7-9 Transformer mounted on central unit. Notice the terminals are marked.

about 24 volts. Low-voltage thermostat cable (#18) might have five or six color-coded wires, depending on the units and the thermostats.

Low-voltage wiring such as thermostats and doorbells do not require outlet boxes. The cable is simply pulled through a hole at the desired location. Connections are clearly marked on both the transformer and the back of the thermostat (FIG. 7-10) with color-coded letters such as W, Y, R, and G (white, yellow, red, and green). Typically, the W connection will affect the heating relay, the Y will control the cooling contactor; the R will connect to the manual fan; and the G will operate the fan automatically in either heat or cool.

Most thermostats have a small mercury-filled glass bulb attached to the bonded strip (FIG. 7-11). This mercury makes the electrical connections, so the base of the thermostat must be mounted level. Use a level and mark the spot for the mounting holes with a pencil. Use a small drill bit to make the holes. (In drywall, screw anchors might be necessary.) Then mount the base on the wall (FIG. 7-12) and snap the cover in place (FIG. 7-13).

SMOKE DETECTORS

Most city ordinances now require smoke detectors to be installed in new construction. Often smoke detectors must be installed in each bedroom as well as in hallways and near stairways. These home fire-alarm devices are about the size of a small light fixture. They are available in battery-operated or direct-wired models. The direct-wired detectors are permanently wired into one of the house circuits. The battery-operated detectors operate independently of the home electrical system and are normally powered by a single, 9-volt alkaline battery that should be replaced every year. Most models will beep periodically to signal that the battery is getting weak.

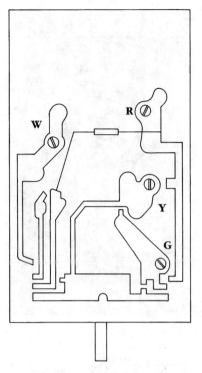

7-10 Marked terminals on back of thermostat.

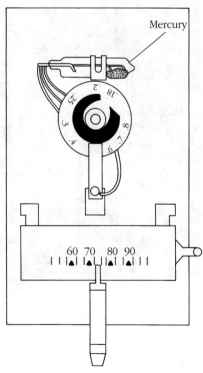

7-11 Thermostat showing mercury switch.

7-12 Level and mount the base of the thermostat to the wall.

7-13 Completed thermostat with cover in place.

Most detectors are mounted on a wall between 6 to 12 inches from the ceiling. Never mount them more than 12 inches from the ceiling and never mount them in corners. Also, do not mount the detector in a path of ventilation where the air flows past the detector faster than in other areas of the room.

The battery-type detector is self-contained and does not require any wiring. Simply mount the bracket using screws, and anchors if necessary, then twist the detector in place (FIG. 7-14). Install the 120-volt models on a box as you would a light fixture (FIG. 7-15). Use a circuit feeding a room where you would notice quickly if a breaker trips (a bathroom for example) because the power must be on all the time. As with the light fixture, connect white wires to white and black wires to black using wire nuts.

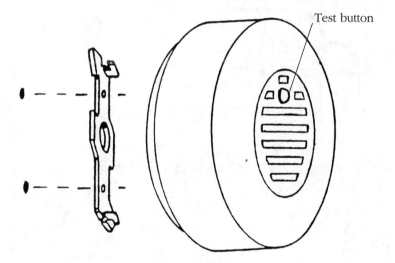

Test button

7-14 Battery-operated smoke detector and mounting bracket.

TELEPHONES

Voltages powering a home telephone service can vary from 48 volts dc to as high as 105 volts ac. Normally, these voltages are not dangerous, but you should take some precautions. Anyone with a pacemaker should not work on a telephone system connected to the service.

Telephone service enters the house by one of two methods. One is through an overhead service connected to a nearby telephone pole (FIG. 7-16), and the other is through an underground service (FIG. 7-17). The telephone company brings the service to the house where, unless they wire the house, the wires will end in a plastic or metal box, normally referred to as a *terminal box*. Because telephone wires are so small, special wire strippers are usually required. An installation kit called "Add-an-Outlet" is

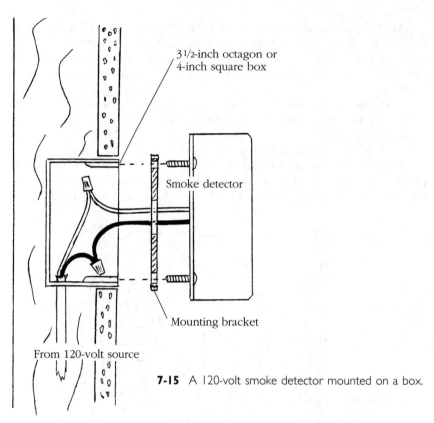

3½-inch octagon or
4-inch square box

Smoke detector

Mounting bracket

From 120-volt source

7-15 A 120-volt smoke detector mounted on a box.

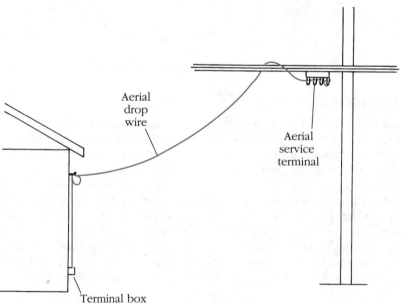

Aerial
drop
wire

Aerial
service
terminal

Terminal box

7-16 Overhead telephone service to a residence.

130 Installing special circuits

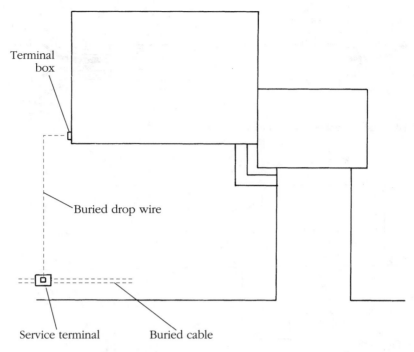

Terminal box

Buried drop wire

Service terminal   Buried cable

7-17 Underground telephone service to a residence.

available from AT&T Phone Centers for about $25. It contains 50 feet of cable and all of the necessary hardware to install three phone jacks. The kit comes with instructions and with the special wire strippers.

The telephone requires two wires to operate. An additional pair is used for a lighted dial or a touch pad. If the cable you install has six wires, the last pair can be used for a separate number or simply folded aside to be used as a spare. The wires should be at least a #24.

A standard cable will have four wires colored yellow, black, green, and red (FIG. 7-18). These wires are connected at the terminal box (FIG. 7-19) and run to the individual jacks either in series (FIG. 7-20) or separately (FIG. 7-21). Running the wires separately would be useful if additional phone numbers are desired.

It is much easier to install the wiring in the walls while the framing is being done, if possible (FIG. 7-22). If not, run the cable in a crawl space under the floor (FIG. 7-23) or along the inside of the fascia of the roof over-hang (FIG. 7-24). Often the cable is simply stapled to the outside wall and brought into the house at the desired location (FIG. 7-25).

Once the cable is inside, try to run it in an inconspicuous manner along baseboards (FIG. 7-26). Use a small wire hook to help fish the cable through a wall (FIG. 7-27).

Wall jacks come in a variety of models (FIG. 7-28), but the connections are pretty much the same inside. The terminals are marked for the color code (FIG. 7-29). Just follow the color codes to make your connections.

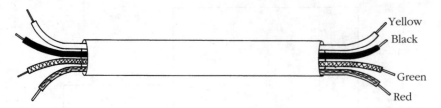

7-18 Typical telephone cable containing four wires.

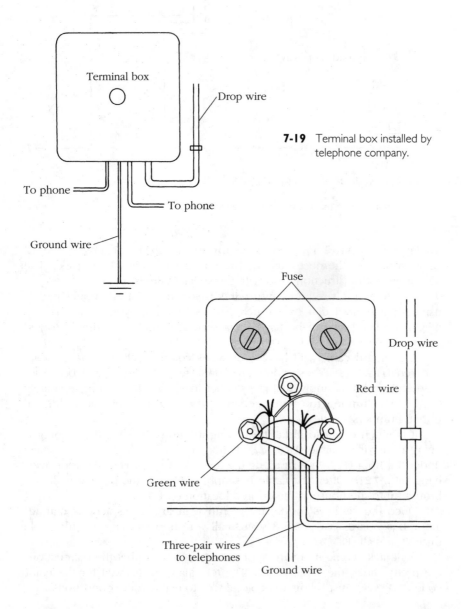

7-19 Terminal box installed by telephone company.

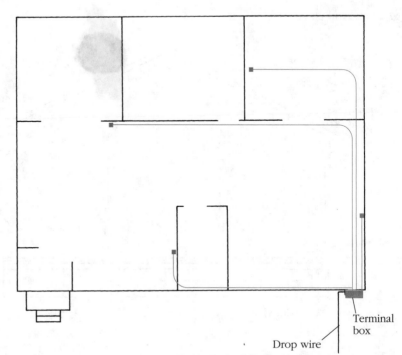

7-20 Series wired phone
jacks.

Terminal
box

Drop wire

7-21 Phone jacks wired
directly to terminal box.

Terminal
box

Drop wire

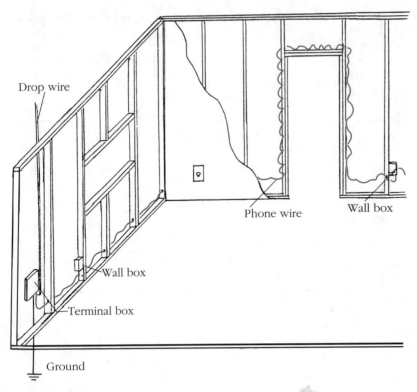

Drop wire

Phone wire

Wall box

Wall box

Terminal box

Ground

7-22 Running phone cable inside the walls of a residence.

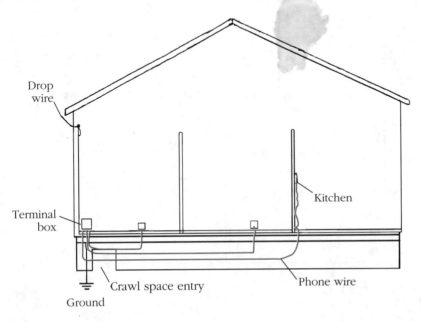

Drop wire

Kitchen

Terminal box

Crawl space entry

Ground

Phone wire

7-23 Run phone cable in the crawl space to the desired locations.

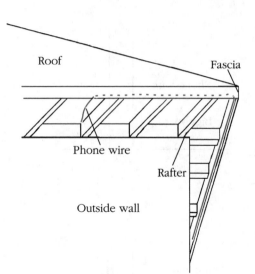

7-24 Run phone cable outside of the house on the underside edge of the roof overhang.

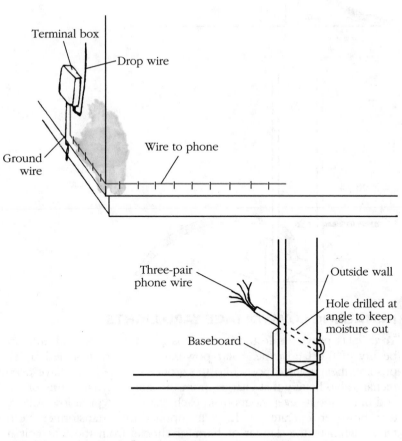

7-25 Staple the phone cable along the outside wall to the desired location.

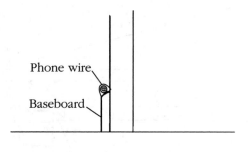

7-26 Phone cable running along the top of the baseboard and hidden by carpet along the bottom of the baseboard.

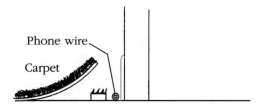

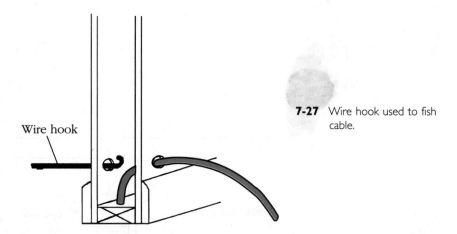

7-27 Wire hook used to fish cable.

INSTALLING LOW-VOLTAGE YARD LIGHTS

Yard lights provide safe travel on steps and walkways, and enhance the beauty of the landscape. Solar powered lights are installed by simply pushing their pointed posts into the ground. These lights have a built-in rechargeable battery that charges from the sun. The light turns on at dark and burns about seven hours from each charge. A typical low-voltage system, however, operates on 12 volts supplied by a transformer. Both systems eliminate the danger of harmful shocks from the 120 volt house system.

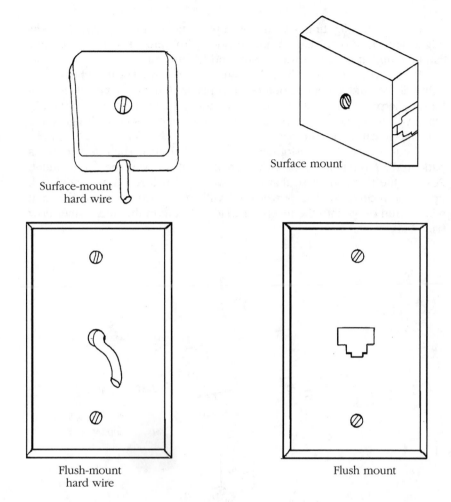

Surface mount

Surface-mount
hard wire

Flush-mount
hard wire

Flush mount

7-28 Surface-mount and flush-mount wall jacks.

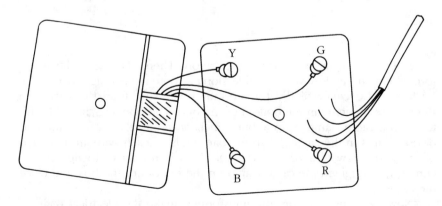

7-29 Marked terminals inside jack.

Low-voltage lighting kits are usually operated by a timer. The kits might have 4, 6, or 10 lights and are priced from about $30 to $60. They can be found at most hardware stores and home centers.

Before you buy your light kit, study your yard. Locate where each light will be, and a suitable outdoor receptacle. Outdoor receptacles must be weatherproof and protected by a ground-fault circuit interrupter (GFCI). Once you have the system planned, buy a kit that has enough lights and low-voltage cable for your needs.

Try to locate the transformer in a protected area, such as beneath a patio roof. Mount the transformer on the outside wall near the outlet. Position the transformer so that the power cord will run down then back up to the receptacle. The power cord will form a drip loop so that rain will not run down into the receptacle. Do not plug in the transformer now (FIG. 7-30).

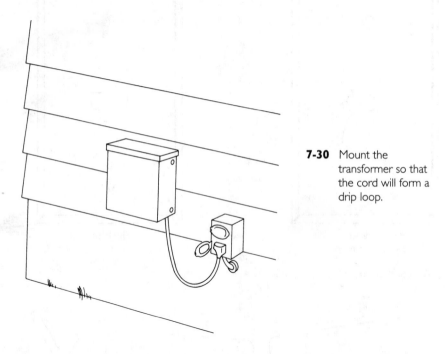

7-30 Mount the transformer so that the cord will form a drip loop.

Place each light in the desired location. Then, starting at the transformer, lay out the low-voltage cable along a route to each light (FIG. 7-31).

Open the cover of the transformer box. Now feed the cable up through the opening in the bottom of the box. The opening should be fitted with a cable clamp. Strip about ¾ inch of the insulation from the end of both wires in the cable. Wrap the bare ends of each wire around each of the two screw terminals (FIG. 7-32). Use a screwdriver to tighten the connections. Tighten the cable clamp on the bottom of the box and close the cover.

Now, working out from the transformer, make the electrical connections according to the manufacturer's instructions. Some connectors have

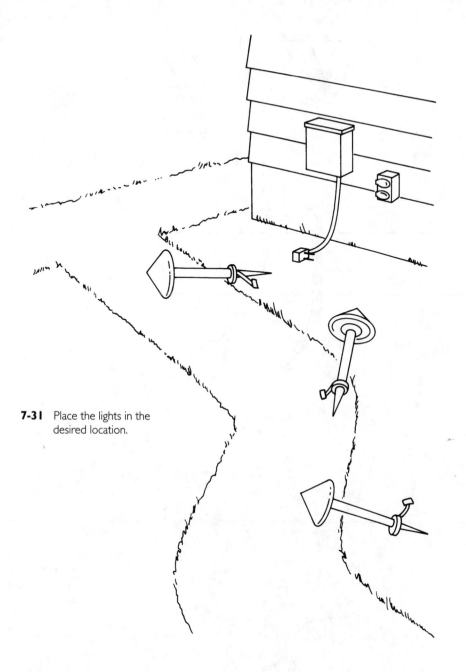

7-31 Place the lights in the desired location.

protruding metal teeth. When the connector is pressed in place, the teeth pierce the insulation and make the connection to the wire (FIG. 7-33).

You also could use wire nuts to make the connection. Cut the cable at the desired location. Then strip about ½ inch of insulation from the cable wires and both light wires. Place the bare ends of two cable wires and one light wire together and screw on a wire nut. Tighten the wire nut

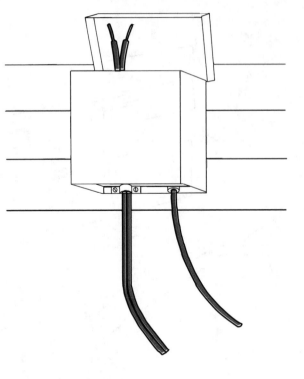

7-32 Connect the wires to the transformer.

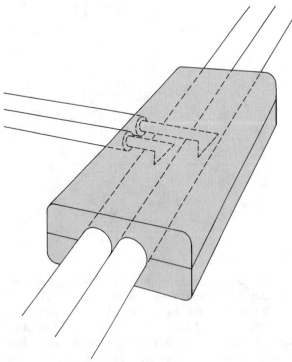

7-33 Some connections have metal teeth that pierce the insulation to make the connection.

clockwise. Connect the remaining wire to the light the same way. Use silicone sealant to weatherproof the connections (FIG. 7-34). Position the lights and push the mounting stakes into the ground.

To bury the cable, press a flat spade several inches into the ground with your foot. Rock the spade back and forth until a small gap is opened. Continue working until you have a narrow slit between each light. Place the cable into the slit and use your foot to press the slit closed. Now plug the transformer into the receptacle, set the timer, and check the installation.

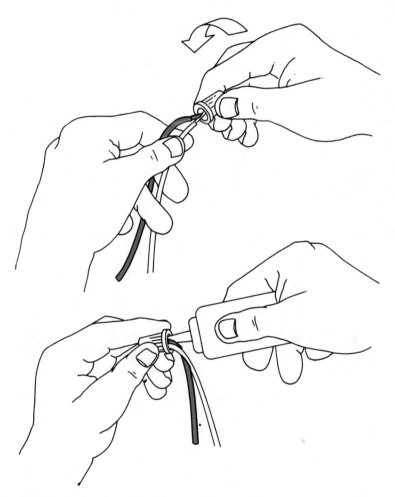

7-34 Silicone sealant will weatherproof the connection.

Chapter 8

Using electrical metallic tubing

Several types of electrical metallic tubing (EMT) are available for electrical wiring. These include *rigid conduit, rigid nonmetallic* (PVC), *EMT* (thin-wall) conduit, *flexible metallic*, and *liquid-tight flexible* metal conduit. Because of the cost, most home wiring limits the use of conduit to the service entrance equipment and connections to various appliances.

Home wiring normally uses rigid, thin-wall, and a flexible metal conduit (FIG. 8-1). Rigid and thin-wall conduit is available in 10-foot lengths and in sizes from ½ to 4 inches. Also available are small lengths that are factory-bent for 90-degree turns. Thin-wall conduit does not use threaded connectors, while rigid conduit comes with the ends threaded (FIG. 8-2). Connectors for thin-wall conduit are designed for use indoors (FIG. 8-3) or for rain-tight connectors at outdoor locations (FIG. 8-4).

Small, exposed runs from boxes to appliances, such as water heaters and air conditioners, normally use flexible conduit. If installed in a dry location, flexible metallic conduit can be used. If the location is exposed to weather, use liquid-tight flexible metal conduit (FIG. 8-5). Connectors are available for both types (FIG. 8-6).

CUTTING

When cutting conduit, use a fine-toothed hacksaw or a tubing cutter. Always ream out the cut ends to remove burrs and sharp edges. This is a must; otherwise, these razor sharp edges will almost certainly strip bits of insulation from the wires as they are pulled through.

Wires being pulled through conduit are restricted by the number of bends in the conduit. A run of conduit between boxes must not have a total of bends that equal 360 degrees (FIG. 8-7). Normally, you won't have

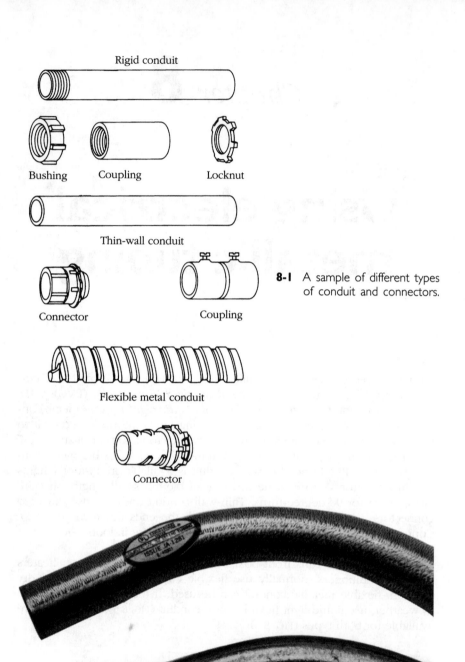

Rigid conduit

Bushing   Coupling   Locknut

Thin-wall conduit

Connector   Coupling

8-1 A sample of different types of conduit and connectors.

Flexible metal conduit

Connector

8-2 EMT (thin-wall) conduit and threaded rigid conduit.

8-3 Thin-wall conduit with coupling and a connector.

8-4 Thin-wall conduit with rain-tight coupling and connector.

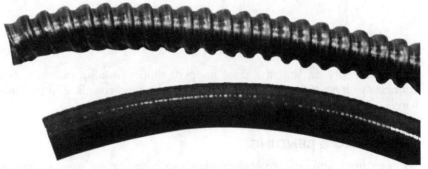

8-5 Flexible metallic and liquid-tight flexible metal conduit.

8-6 Connectors used with flexible conduit.

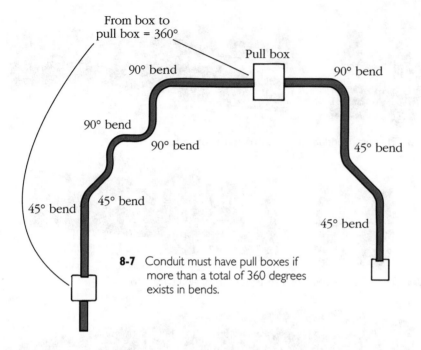

From box to
pull box = 360°

Pull box

90° bend

90° bend

90° bend

90° bend

90° bend

45° bend

45° bend

45° bend

45° bend

45° bend

8-7 Conduit must have pull boxes if more than a total of 360 degrees exists in bends.

such a case anyway, but if it does happen, simply install a pull box. No connections have to be made inside; it's just a convenient place to pull the wire.

MEASURING & BENDING

To keep from kinking thin-wall conduit when bending it, use a conduit bender (FIG. 8-8), which you can purchase or rent. The conduit bender is marked in degrees to help you make the bend desired. It is usually better to bend the conduit first, then cut the necessary length. When the handle of the bender is straight up, it makes a 45-degree bend (FIG. 8-9), and if the bend is continued until the stub end is vertical, it makes a 90-degree bend (FIG. 8-10).

Often you might want to bend a stub end to a desired length. For example, if you want a 90-degree bend extending 12 inches, begin by placing the bender on the conduit at a specific location. This is a point equal to the length of the stub end minus an amount used to make the bend called the *take-up*. The take-up is 5 inches for ½-inch conduit, 6 inches for ¾-inch conduit and 8 inches for 1-inch conduit.

For a 12-inch stub using ½-inch conduit, subtract the 5-inch take-up for a remainder of 7 inches. Set the mark on the bender 7 inches from the end (FIG. 8-11). Begin the bend by applying foot pressure on the step and pulling steadily on the handle until the stub end is vertical (FIG. 8-12). For back-to-back bends, make the first bend, then measure the conduit for the

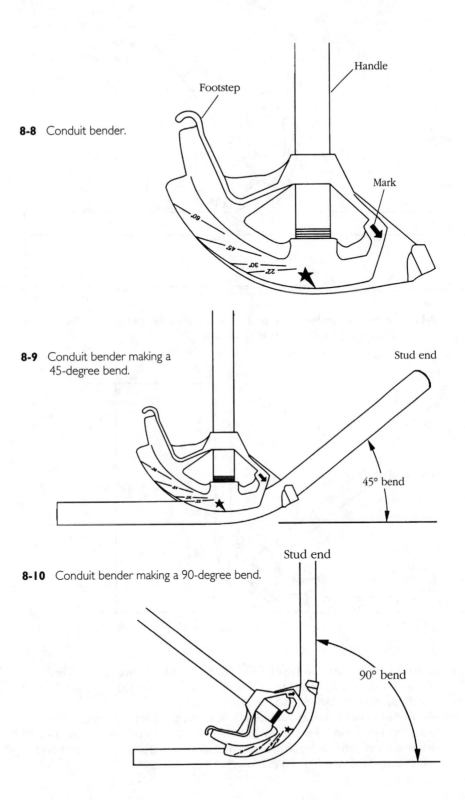

8-8 Conduit bender.

Footstep

Handle

Mark

8-9 Conduit bender making a
45-degree bend.

Stud end

45° bend

8-10 Conduit bender making a 90-degree bend.

Stud end

90° bend

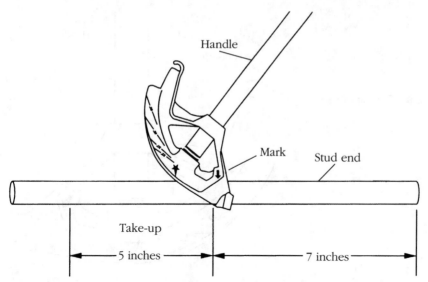

8-11 Use the mark on bender to set the bender in the desired location. For instance, 7 inches from the end will make a 12-inch stub.

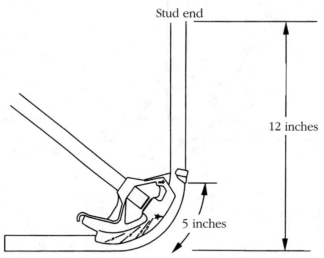

8-12 Conduit bender uses 5 inches to make a bend in ½-inch conduit.

desired length. Mark the bender for the back-to-back overshoot. This usually is 2¼ inches for ½-inch thin-wall conduit. Add this to the desired length, then make the second bend (FIG. 8-13).

Use an offset bend, or kick bend, to make a straight-in connection to a box. This bend can be made by first bending the conduit up about 5 to 10 degrees, then turning the conduit over, moving it forward in the bender, and making a second bend (FIG. 8-14).

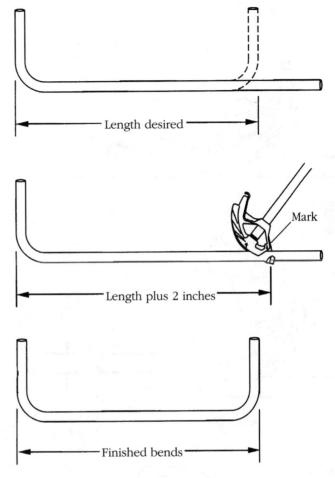

8-13 Make back-to-back bends by adding 2¼ inches to the desired length of ½-inch conduit.

INSTALLING & PULLING THE WIRE

Conduit runs must be supported at least every 10 feet and within 3 feet of each outlet. After the conduit is in place, pull the individual wires. When conduit is properly installed, using metal boxes and with all connections solid and tight, use the conduit for the equipment ground (the bare ground wire). This means that only two wires are necessary for 120-volt and 240-volt circuits, and three wires are necessary for the 120/240-volt circuits. If there is a loose connection anywhere in the conduit, however, the system will not be properly grounded. The wires must run continuously from box to box without splices. Connections are to be made in the boxes with no substitutions for color coding.

For short runs, the wire can usually be pushed from box to box, but longer runs with bends will probably require the use of a fish tape (FIG.

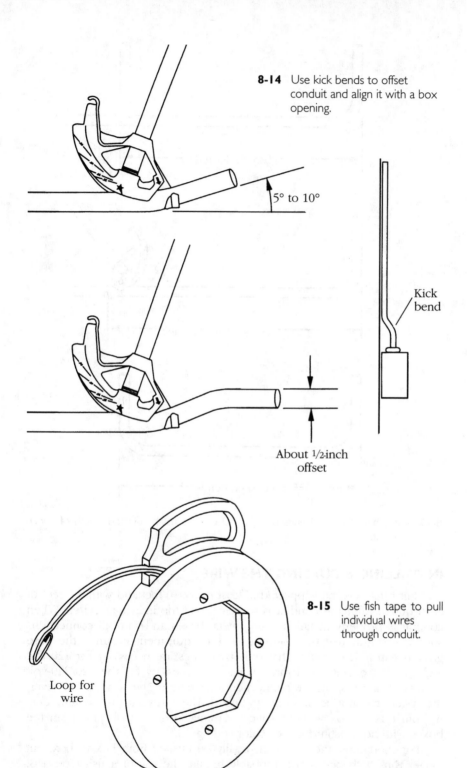

8-14 Use kick bends to offset conduit and align it with a box opening.

5° to 10°

Kick bend

About ½-inch offset

8-15 Use fish tape to pull individual wires through conduit.

Loop for wire

8-15). Fish tape is a flexible steel tape about ⅛ inch wide, 50 to 100 feet long, with a loop in the end. The fish tape is inserted through the conduit to the desired box. Next, the wires are connected as shown in FIG. 8-16. The fish tape is then pulled, pulling the wires through the conduit (FIG. 8-17).

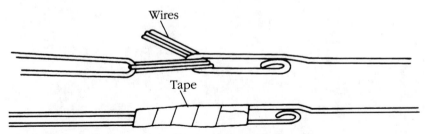

8-16 Strip wire ends of insulation, thread them through the loop, and bend them back. Use plastic tape to keep the wires together for a smoother pull.

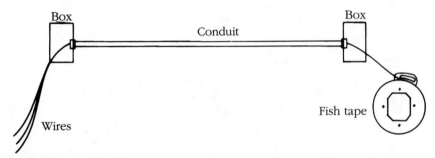

8-17 Fish tape pulling wire through conduit.

Remember that no color-code substitutions are allowed when using conduit, as with nonmetallic sheathed cable. Switch loops should have only the black or red wires and should not use the white wire marked with black tape. Otherwise the wiring is connected the same except the ground wire is not needed.

Chapter **9**

Electrical service for rural areas

Rural wiring is essentially the same as electrical systems in the city. One major difference, however, is the type of material used to install an electrical service in wet and corrosive areas such as those that house livestock. Conduit is rarely used because the high humidity and the corrosive action of animal excretion will cause metal boxes and conduit to rust away. If NM-type nonmetallic sheathed cable (the kind used for dry locations) is installed, mildew and rot attack the cable from the outside, while the jute filler pulls moisture inside.

To avoid this, type NMC cable was developed—its individual conductors are embedded in a plastic jacket. Another option is to use UF cable, which is more expensive, but which can be used where types NM and NMC cable are used as well as underground. These wet and corrosive conditions also dictate that nonmetallic outlet boxes be used.

The other difference between rural and city wiring is that most rural electrical services use a meter pole centrally installed in the yard (FIG. 9-1). The pole's location reduces the length of cable necessary to provide service to two or more buildings. Normally, the utility company installs the pole and provides service to the pole, but from the top of the pole down is usually the owner's responsibility.

CONSIDERATIONS FOR RURAL SERVICE

The mechanical strength of wires is an important consideration for planning rural electrical service because the wires are run overhead, and are therefore subject to damage by high winds and ice. The Code requires that spans up to 50 feet must use at least #10 copper wires; #8 must be

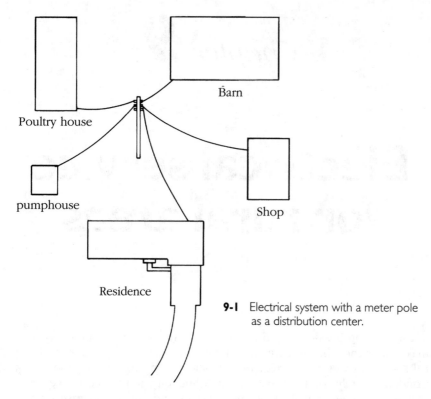

Poultry house

Barn

pumphouse

Shop

Residence

9-1 Electrical system with a meter pole as a distribution center.

used for longer spans. If aluminum wire is used, the next larger size must be selected.

Another important consideration is the length of run. Voltage drops as the distance increases; for example, a 120-volt circuit carrying 15 amps and using #10 wire should extend no farther than about 70 feet. This will ensure that the voltage will drop no more than 2 percent over the distance of the run.

When determining the size of a rural electrical system, you must calculate the amount of power that will be used. In estimating the size of the service, the residence might need 150 amps; a shop with a welder and other equipment, about 75 amps; a poultry house, about 40 amps; a barn, about 30 amps; and a pump house, about 10 amps. The service entrance panels and wire size to each location would be a 150-amp service and 2/0 wire for the residence; a 150-amp service and 2/0 wire for the shop; a 50-amp service and #3 wire for the poultry house; a 60-amp service and #3 wire for the barns; and a single circuit with a 30-amp service and #6 wire for the pump house.

The wire sizes are based on a distance up to 100 feet. For up to 150 feet, the next larger size is used. Other factors that determine the wire size include the type of wire, whether it is in conduit, and if it is buried or run overhead. Check with your local building inspector because weather conditions in some areas can make a difference in the installation.

INSTALLING POWER POLES

It is important to check with your utility company to know what their requirements are. Often they will help in the planning. If you plan to have service only to one outbuilding, you might be able to run a circuit from the house service entrance panel. For two or more buildings, a meter pole is normally installed at a central distribution point. Each building must have a disconnect and a service entrance panel just like the residence, except that no meter is installed.

Wiring the meter pole is best accomplished before the pole is erected. This means mounting the conduit, meter base, and any disconnect switch, and installing the wiring, including the ground wire. Then the pole is placed in the ground. Install the insulator rack or individual insulators (FIG. 9-2) near the top of the pole. Mount one rack for the conductors from the utility company and one for the building served from the pole (FIG. 9-3). Attach the entrance head above the insulators so that the wires will form drip loops (FIG. 9-4). Mount the meter base so its bottom is about 5 feet above ground level. If only one run of conduit is used, a disconnect switch is not allowed.

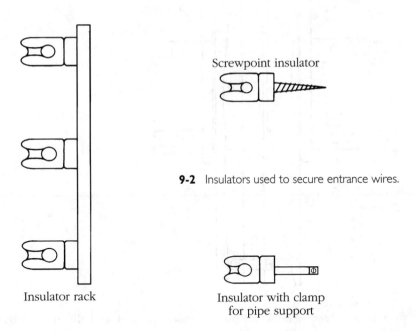

Screwpoint insulator

9-2 Insulators used to secure entrance wires.

Insulator rack

Insulator with clamp
for pipe support

Install two hot wires from the top terminals of the meter base through the entrance for the incoming service. Run two more hot wires from the bottom terminals of the meter base through the service head for the service to the building. Install a neutral wire from the center terminal of the meter base through the service head for the neutral connection. Another neutral wire is not needed because the neutral to the building is made by

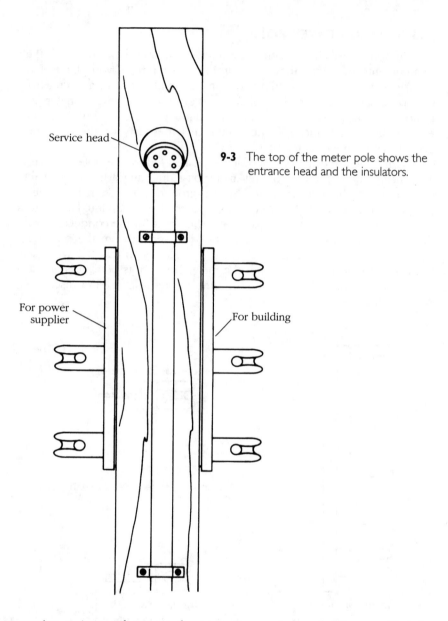

Service head

9-3 The top of the meter pole shows the entrance head and the insulators.

For power supplier

For building

running a *jumper* between the incoming neutral and the neutral to the building (FIG. 9-5). Connect the grounding wire, usually a #6, from the utility company's neutral to a ground rod (FIG. 9-6). The copper ground rod is at least ½ inch in diameter and 8 feet in length.

If a disconnect switch is desired, wire the meter pole using two runs of conduit. In FIG. 9-7, two hot wires and a neutral run from the incoming service through the meter base and disconnect (if installed), then through the entrance head to the building. The connections to these large wires are made with split-bolt connectors (FIG. 9-8). The wires are connected as

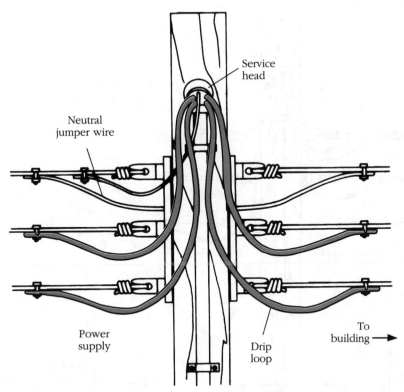

Service head

Neutral jumper wire

Power supply

Drip loop

To building ⟶

9-4 The entrance head is mounted above the insulators so the wires form drip loops.

shown in FIG. 9-9, then wrapped with electrical tape to a thickness equal to the insulation of the wires.

The disconnect switch should be a double-pole, double-throw type so an emergency generator can be installed. It should be wired in the manner shown in FIG. 9-10.

OVERHEAD & UNDERGROUND SERVICE TO BUILDINGS

If two buildings are close together and maybe the second building only needs 120 volts, simply tap off one hot wire and the neutral as shown in FIG. 9-11. There are a number of ways to bring wires into a building. You can use a special entrance head (FIG. 9-12A), or you can use an ordinary entrance head with a short length of conduit and an outlet box (FIG. 9-12B). Install the service entrance for each building the same as the residential service entrance, except do not install a meter.

An underground service to another building might be desirable. If the service has only two circuits and is protected by a 30-amp breaker, install the service as shown in FIG. 9-13.

To extend service in a long building and avoid the expensive circuit runs, a subpanel (FIG. 9-14) is often used. The subpanel acts as a branch

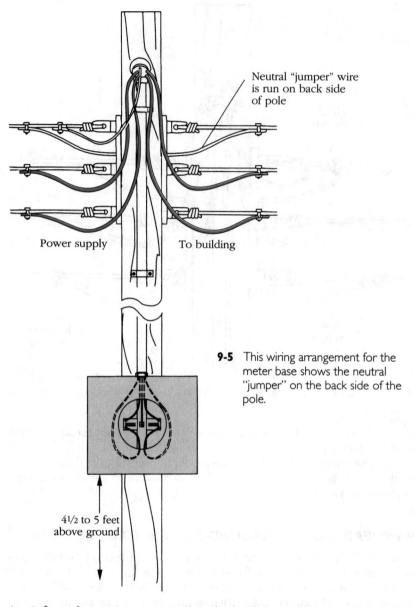

Neutral "jumper" wire
is run on back side
of pole

Power supply   To building

9-5 This wiring arrangement for the
meter base shows the neutral
"jumper" on the back side of the
pole.

4½ to 5 feet
above ground

circuit from the service entrance panel where the wires are connected to a
circuit breaker just like any 120/240-volt circuit. The difference with the
subpanel is that only the bare ground wires are grounded with the cabinet. Connect the neutral wires to a floating neutral busbar (FIG. 9-15) and
ground them only on the service entrance panel. Do not ground the neutral busbar in the subpanel to the cabinet. Then wire the individual circuits
into the subpanel as you would the service entrance panel.

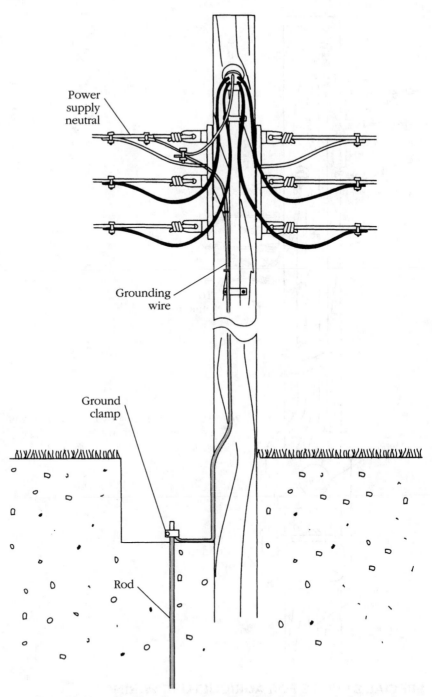

9-6 Connect the grounding wire to the utility company's neutral, then run it to a ground rod.

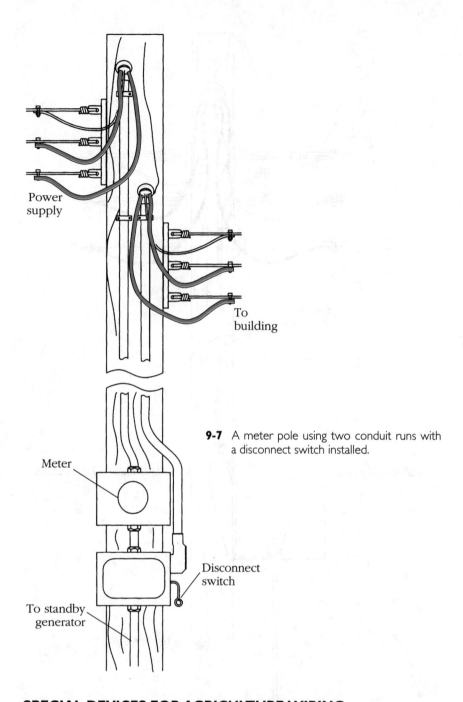

Power
supply

To
building

9-7 A meter pole using two conduit runs with a disconnect switch installed.

Meter

Disconnect
switch

To standby
generator

SPECIAL DEVICES FOR AGRICULTURE WIRING

Normally the cables are surface-mounted, both to allow for routine inspection and maintenance, as well as to reduce damage from rodents. In surface wiring, holes are not drilled through wood members. The cable is

9-8 A split-bolt solderless connector.

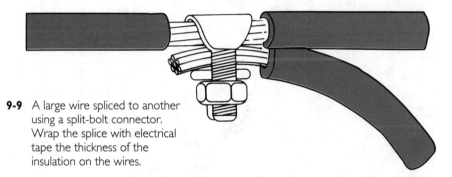

9-9 A large wire spliced to another using a split-bolt connector. Wrap the splice with electrical tape the thickness of the insulation on the wires.

supported by nonmetallic cable straps secured by corrosion-resistant nails. These nails should be located at 2-foot intervals on horizontal runs, at 3-foot intervals on vertical runs, and within at least 8 inches from each box.

Where the wiring is subject to damage, conduit might be desirable. You can use type 80, heavy-walled, rigid PVC conduit. Install it with weatherproof receptacles and switches (FIG. 9-16 and FIG. 9-17). These boxes have threaded openings (FIG. 9-18) for rain-tight fittings. Use non-metallic flexible tubing to feed equipment, and use special connectors (FIG. 9-19) to connect the tubing to the boxes. These connectors use an O-ring (FIG. 9-20) to ensure a proper seal. Light fixtures that might be exposed to water must be watertight, and those exposed to physical damage must also be protected by a guard (FIG. 9-21).

LOW-VOLTAGE RELAYS FOR REMOTE CONTROLS

Sometimes it is desirable to control lights or equipment from several locations. An inexpensive way to do this is to install a low-voltage (24 volts) remote control system. The system operates from a low-voltage transformer. In FIG. 9-22, 120 volts is stepped down to 24 volts.

The cable is usually #18 three-wire conductor for outdoor use (FIG. 9-23). This cable has a rib on one edge to identify the common conductor. This wire must be connected to the same numbered terminal on each

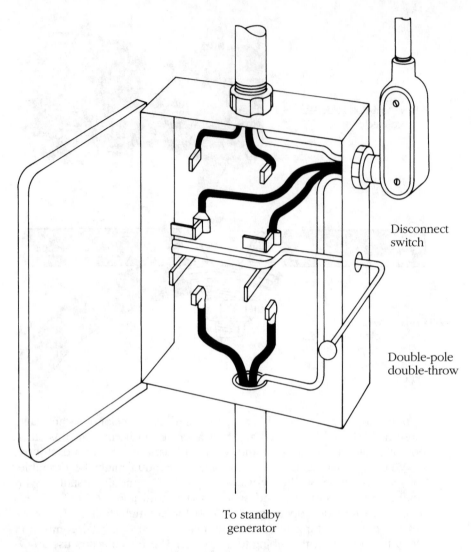

Disconnect
switch

Double-pole
double-throw

To standby
generator

9-10 A double-pole, double-throw disconnect switch.

switch. The remaining components needed are the relay and switches
(FIG. 9-24).

Mount the transformer in a central location to serve the remote-controlled circuits. Install a low-voltage relay at each light to be controlled.
Connect the two wires from the relay just like a switch—one to the black
wire from the source and the other to the black wire to the light. The
other end of the relay has three connections for the three-wire, low-voltage cable. Next, run the cable from the transformer to the relay and then
to the switches used to control the light (FIG. 9-25).

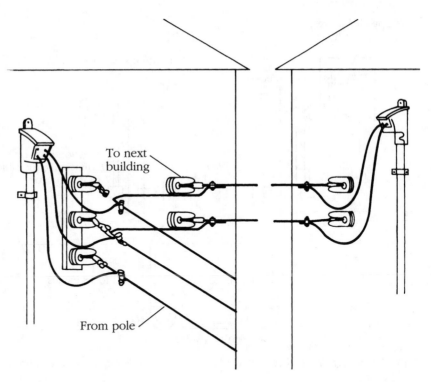

9-11 Run a service to a second building by tapping into the wires serving the first building.

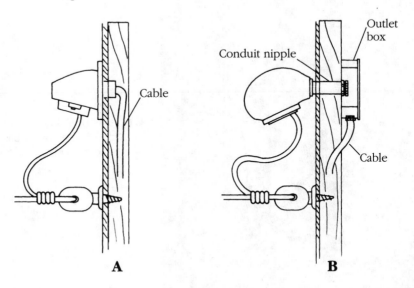

A **B**

9-12 Two methods of bringing wires into a building: (A) a special entrance head; (B) a regular entrance head and conduit.

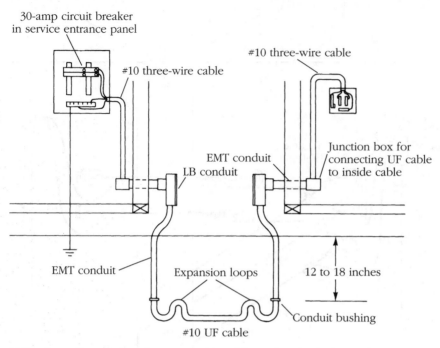

30-amp circuit breaker
in service entrance panel

#10 three-wire cable

#10 three-wire cable

Junction box for
connecting UF cable
to inside cable

EMT conduit

LB conduit

EMT conduit

Expansion loops

12 to 18 inches

Conduit bushing

#10 UF cable

9-13 An underground service protected by a 30-amp breaker in the service entrance panel.

9-14 A subpanel with a floating neutral busbar and grounding bus bar.

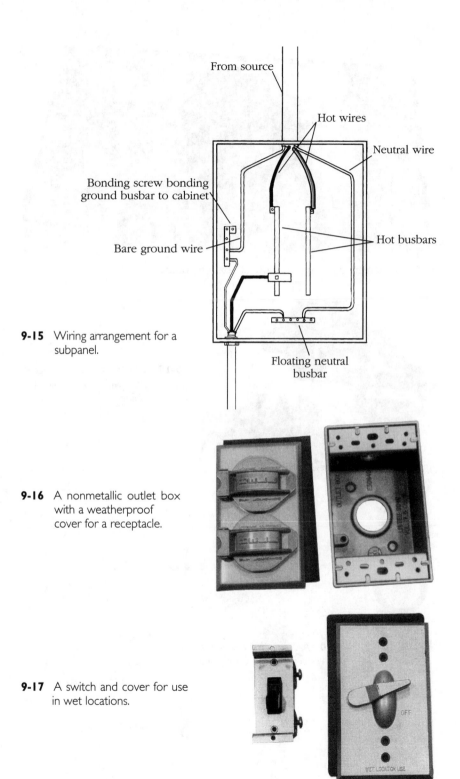

From source

Hot wires

Neutral wire

Bonding screw bonding
ground busbar to cabinet

Bare ground wire

Hot busbars

9-15 Wiring arrangement for a
subpanel.

Floating neutral
busbar

9-16 A nonmetallic outlet box
with a weatherproof
cover for a receptacle.

9-17 A switch and cover for use
in wet locations.

9-18 The threaded opening in an outlet box used outdoors.

9-19 Weatherproof connectors used with nonmetallic flex tubing.

9-20 A connector disassembled to show the O-ring seal.

9-21 A watertight light fixture with a protective guard.

9-22 A low-voltage transformer with a 24-volt output.

24 V

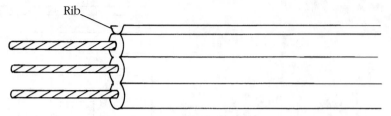

9-23 Three-wire, low-voltage cable with a rib to identify the common conductor.

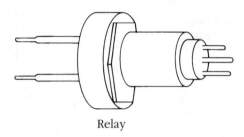

Relay

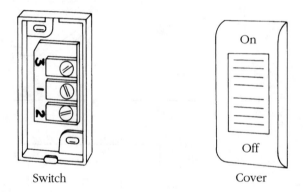

Switch   Cover

9-24 A 24-volt relay and three-terminal switch used in remote-controlled system.

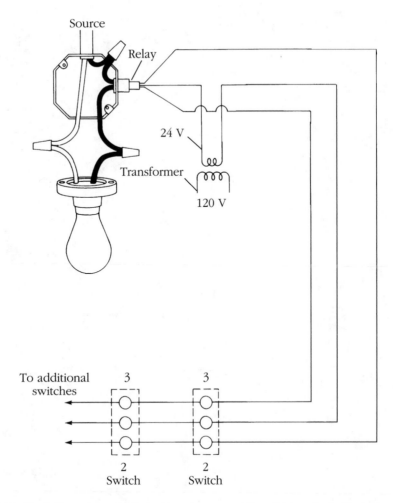

9-25 Typical remote-controlled wiring arrangement for a light controlled by two switches.

Chapter **10**

Easy repairs
you can do

A do-it-yourselfer can correct a variety of simple electrical problems. The important thing to remember is to turn the power off before beginning any repair.

REPLACING ELECTRICAL CORDS

Electrical cords are the most common source of problems for small appliances because the cords are often abused when they are unplugged. Furniture pushed too close to the wall also can damage a cord. Try to avoid the conditions shown in FIG. 10-1, and you will greatly extend the life of an electrical cord.

A cord should not overheat, but if the cord is too small or has a loose connection, the heat—and time—will cause the insulation to become brittle and cracked. The insulation also can become frayed from wear (FIG. 10-2). When either happens, the cord becomes dangerous and should be replaced, rather than repaired.

To replace a cord, you need a small screwdriver, wire strippers, and side-cutting pliers. Always replace the old cord with the right size and type of cord to do the job (FIG. 10-3). For general use around the home, most small appliances (lamps, radios, fans, etc.) use a rubber or plastic insulated lamp cord. These cords come in sizes of #18 and #16. For portable power tools, you can buy a heavier-duty cord that will take more abuse; it comes in sizes #18 to #10. Appliances that heat require a special cord called a *heater cord*, which has an asbestos covering with a braided jacket. Heater cords are available in sizes #18 to #12.

After selecting the type of cord you need, make sure the wire size is at least as large as the old one. Because you must install the new cord

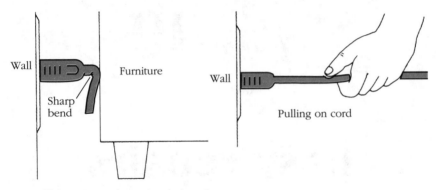

10-1 Things not to do to electrical cords.

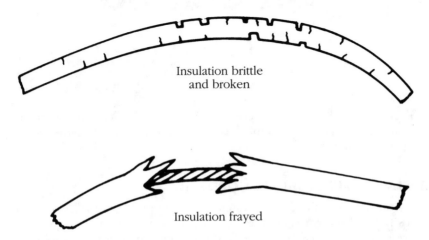

10-2 Dangerous cord conditions that can cause a fire or shock.

exactly the same as the old one, determine what type of strain relief was used in the old cord. This might be in the form of a knot or a clamp where the cord goes into the appliance. The cord will have some slack between this knot or clamp and the connections. Install the new cord the same way. Make the connections using the screw terminals or wire nuts as necessary, observing any color code that might be used.

REPLACING PLUGS

Often the cord will be good and the plug will need replacing. This is quick repair that sometimes is neglected until the plug becomes dangerous. You'll need a small screwdriver, wire strippers, and side-cutting pliers.

If the plug is broken, or if it is the molded kind that cannot be taken apart, replace it with a new one. Some male plugs can be tightened and made to stay in receptacles by spreading split prongs with the tip of a

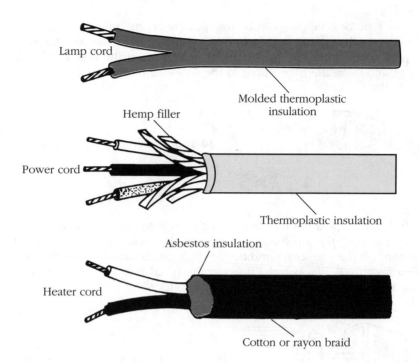

Lamp cord

Molded thermoplastic
insulation

Hemp filler

Power cord

Thermoplastic insulation

Asbestos insulation

Heater cord

Cotton or rayon braid

10-3 Types of replacement cords available.

screwdriver (FIG. 10-4). If the prongs are solid, spread them a little farther apart.

If the plug is a quick-connecting type, press the prongs together, and pull the prong assembly and wire from the cover. Now spread the prongs apart to release the wire (FIG. 10-5).

To install the quick-connecting plug, cut the end of the cord straight across, with no bare wire sticking past the insulation. Stick the wires through the plug cover and then into the opening in the prong assembly. Press the prongs together to make the electrical connection. Now push the prong assembly back into the plug cover (FIG. 10-6).

10-4 Spread the prongs with a screwdriver.

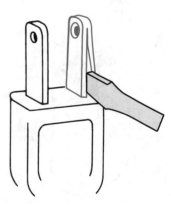

One type of male plug can be installed by prying out an anchor with a small screwdriver, and pulling out the prongs. Thread the cord through the plug cover, place the bare ends of each wire in the notches in the prongs, and make the connections to the screw terminals (FIG. 10-7).

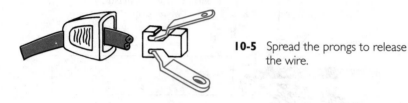

10-5 Spread the prongs to release the wire.

If the faulty plug is a female plug with screw terminals, loosen the top screw. Then pry the top from the bottom with a screwdriver. Thread the cord through the plug cover and tie the Underwriter's knot. Insert the bare ends of each wire through the holes in the metal bars and connect them to the terminal screws. Reassemble the plug (FIG. 10-8).

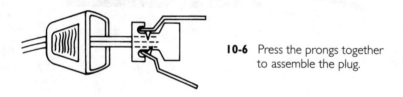

10-6 Press the prongs together to assemble the plug.

If you are dealing with a standard plug, remove the insulating disk to get to the terminal screws. If it is a grounded plug, there will be three screws; if not, there will be only two. Loosen the screws enough to remove the wires (FIG. 10-9).

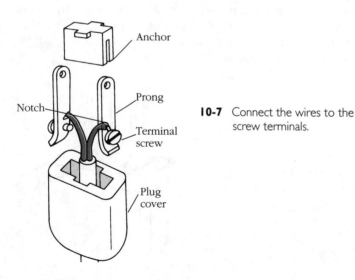

Anchor

Prong

Notch

Terminal screw

Plug cover

10-7 Connect the wires to the screw terminals.

If the plug is a heavy-duty type, there is probably some kind of clamp that grips the back part of the plug to the cord. This protects the connections inside the plug if there is any strain put on the cord. The front part of the plug is often held in place by three screws. Loosen the clamp screw

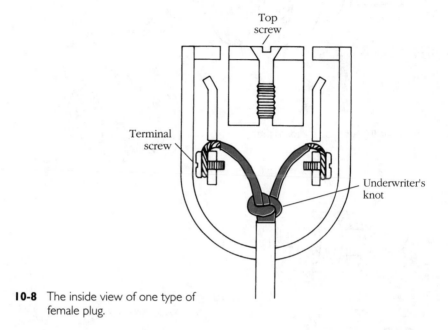

Top screw

Terminal screw

Underwriter's knot

10-8 The inside view of one type of female plug.

and the three screws on the end to separate the plug. Then loosen the terminal screws and remove the wires (FIG. 10-10).

To install a standard plug or heavy duty plug, insert the end of the cord through the back of the plug. Next, remove about 2 inches of the outer insulation. Lamp cords do not have this cover. Separate lamp cords about 2 inches from the ends (FIG. 10-11).

To remove the insulation, strip about ½ inch of insulation from each wire (FIG. 10-12). Twist the strands of each wire so that they will bind together. This keeps them from spreading out when they are tightened under the screws (FIG. 10-13).

If there is no strain-relief clamp, you should tie a knot to take up the strain. Tie the Underwriter's knot as shown in FIG. 10-14. Now tighten the knot so that the free ends are of equal length.

Pull the knot down inside the plug. Bend the bare ends of each wire into a clockwise loop (FIG. 10-15). Now bend the wires around the prongs. Make the connections to the screw terminals. If the plug is a polarized plug, one of the prongs will be smaller. Connect the black wire to this terminal. It should have a brass screw. Connect the white wire to the silver-colored terminal (FIG. 10-16).

Replace the insulated disk if your plug has one. If the plug is the grounded type, connect the black wire to the brass terminal, the white

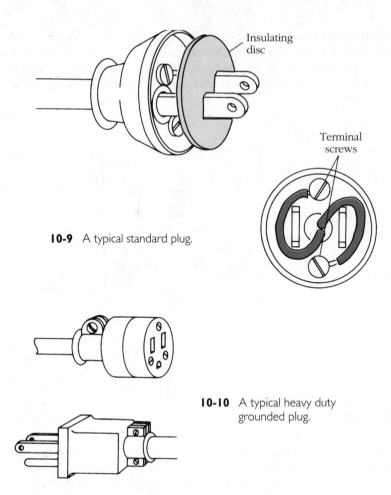

10-9 A typical standard plug.

10-10 A typical heavy duty grounded plug.

wire to the silver-colored terminal, and the green wire to the green termi-nal for the ground prong. Assemble the two parts of the plug and tighten the strain clamp (FIG. 10-17).

REPLACING LAMP SOCKETS

Faulty lamps can be diagnosed by first checking the bulb, then the plug and cord. If these are good, the problem usually can be solved by replac-ing the lamp socket. You will need a small screwdriver, wire strippers, and side-cutting pliers. An adjustable wrench might be necessary to get the lamp apart.

First unplug the lamp. Then remove the shade, shade bracket, and bulb. Examine the outer shell of the socket for a place marked PRESS (it should be near the switch). Hold the cap (the bottom of the socket) steady with one hand while pressing the mark on the shell.

Twist the shell slightly and remove it from the cap (FIG. 10-18). Beneath the outer shell, you should find an insulating sleeve covering the

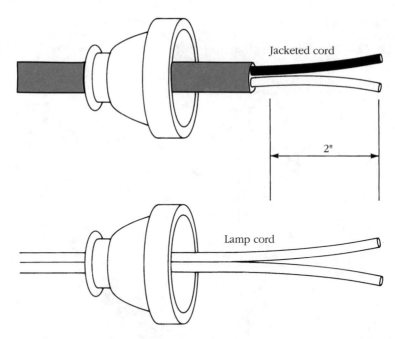

Jacketed cord

2"

Lamp cord

10-11 Separate the wires about 2 inches from the end.

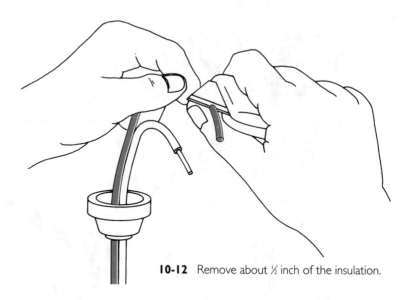

10-12 Remove about ½ inch of the insulation.

socket. Remove the insulating sleeve and you will see the screw terminals (FIG. 10-19).

Use a screwdriver to disconnect the two wires going to the socket and remove the socket, leaving the socket cap in place. The new socket will come with a socket cap, but you usually don't need to replace the old one.

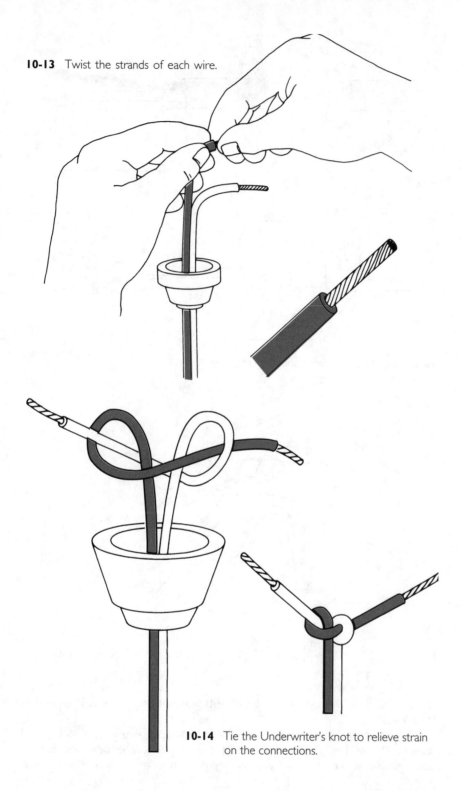

10-13 Twist the strands of each wire.

10-14 Tie the Underwriter's knot to relieve strain on the connections.

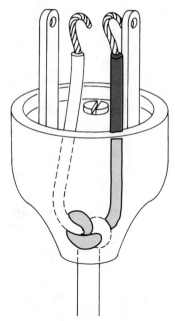

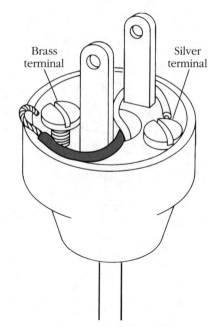

10-15 Form the bare ends into clockwise loops.

10-16 Make the connections to the screw terminals.

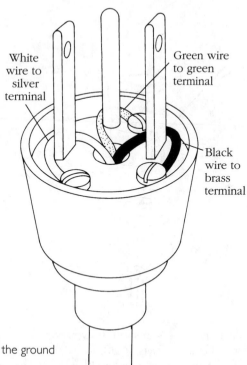

White
wire to
silver
terminal

Green wire
to green
terminal

Black
wire to
brass
terminal

10-17 Connect the green wire to the ground terminal.

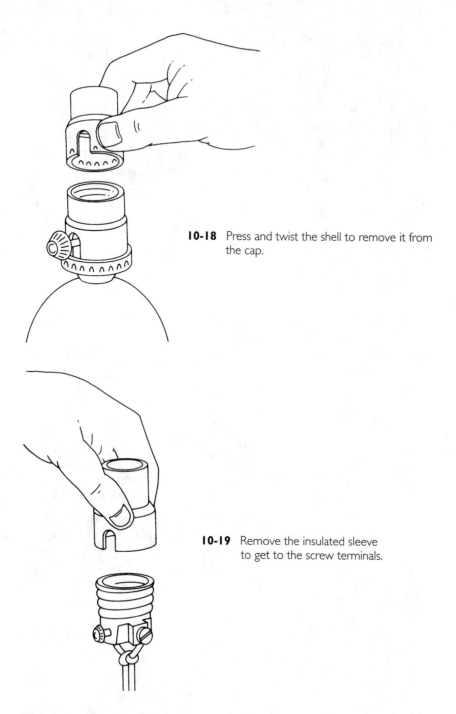

10-18 Press and twist the shell to remove it from the cap.

10-19 Remove the insulated sleeve to get to the screw terminals.

Reconnect the wires to the two screw terminals on the new socket and reinstall the socket (FIG. 10-20).

At this point, you might consider replacing the regular socket with a three-way type (FIG. 10-21). The wire connections are the same—simply

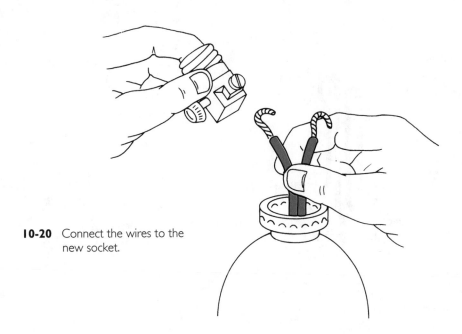

10-20 Connect the wires to the new socket.

remove the old socket, leaving the socket cap and Underwriter's knot in place, and install the new one.

REPLACING SWITCHES

Switches do fail eventually and, regardless of the type, are simple components to replace. Before working on any circuit, though, make sure the power is turned off. Only a few tools are required—a screwdriver, needlenose pliers, and a voltage tester of some kind. A neon test lamp will work fine. Make sure the new switch has the UL stamp showing it has met the minimum safety standards established by Underwriters' Laboratories. If the wiring happens to be aluminum, make sure the switch is stamped CO/ALR on the mounting bracket (FIG. 10-22).

Single-pole switch

For replacing a single-pole switch (the most common switch in the house), begin by removing the screws holding the switch plate, then the switch plate. If you are sure the circuit is dead, remove the screws holding the switch in place (FIG. 10-23). Next, gently pull the switch and slack wires from the box. Loosen the terminal screws, remove the old switch, and install the new one (FIG. 10-24). If the switch has been back-wired, insert a small screwdriver in the slot next to the wire and release the wire. Make sure the switch is right side up by checking the ON/OFF markings. The toggle should be down when the light is off. Fold the slack wires back in the box and mount the new switch to the box. Replace the switch plate and turn the power back on.

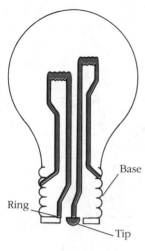

Base

Ring

Tip

10-21 A three-way socket has the same wire connections.

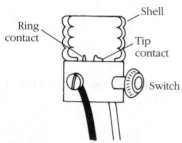

Shell

Ring contact

Tip contact

Switch

10-22 CO/ALR stamped on a switch means copper or aluminum wire can be used.

Three-way switch

Three-way switches are found where you want to turn a light on or off at two different locations—not three, as the term implies. The term *three-way* means that the switch has three terminals for three wires instead of two. There are no ON and OFF markings.

To find which switch is bad, turn the switches until the light is on (FIG. 10-25). Now turn each switch to the opposite position. This sets the switches so you can make the tests. Remember the circuit will be live; be careful (FIG. 10-26).

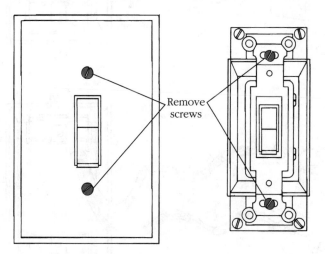

10-23 Two screws hold the switch plate in place, and two screws hold the switch to the switch box.

Remove the switch cover from one of the switches. You should see three terminals. If they are not accessible, you'll have to pull the switch from the box to make the test. In this case, turn off the breaker and remove the switch-mounting screws. Pull the switch out far enough to get to the terminals. The wires will hold the switch clear of the box.

Turn the breaker back on, and touch the end of one probe to one of the terminals. Touch the other probe to one of the other terminals (FIG. 10-27). Now move one of the probes to a different terminal (FIG. 10-28). Now check for voltage between the remaining two terminals (FIG. 10-29). If you found voltage (the test lamp burned) at any one of the combinations, the switch is bad. If the test lamp showed no voltage, the switch is good. **CAUTION!** This does not mean there is no voltage on the switch—There is. You are only checking for a break in the switch.

After locating the faulty switch, turn off the breaker.

Determine which screw is the common terminal and which wire connects to it before disconnecting the old switch. This is the only wire you need to get right. If the screws are on the front of the switch, the common terminal is usually black or copper-colored. If the screws are on the sides,

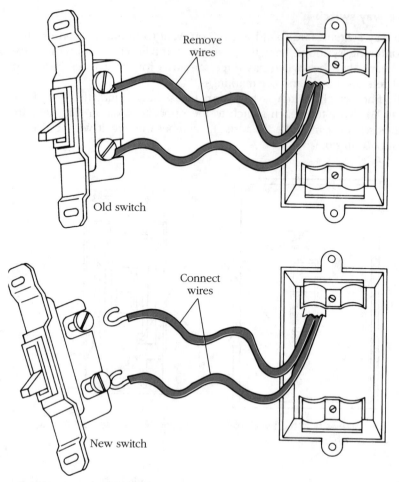

Remove wires

Old switch

Connect wires

New switch

10-24 Replacing a single-pole switch.

the common terminal is usually on one side by itself. If the screws are on the ends, the common terminal is usually on one end by itself (FIG. 10-30). Often, the common terminal is labeled.

After you have located the common terminal, disconnect the wire coming from it and bend the wire away from the switch so that you can identify it later. Disconnect the other two wires and bend them to the other side (FIG. 10-31).

Examine the new switch to locate the common terminal. The box the switch came in should have a wiring diagram. Connect the common wire to the common terminal (FIG. 10-32), and connect the other two wires to the remaining terminals. It doesn't matter which of the last two wires are connected to which terminal—as long as the common is connected properly. Fold the wires back into the box, replace the switch-mounting screws, and then replace the switch cover. Now turn the breaker back on.

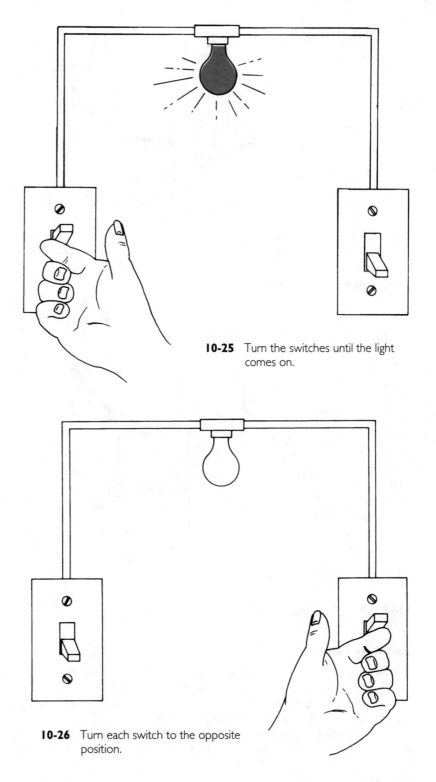

10-25 Turn the switches until the light comes on.

10-26 Turn each switch to the opposite position.

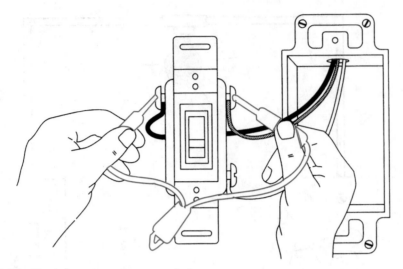

10-27 Check for voltage at two terminals.

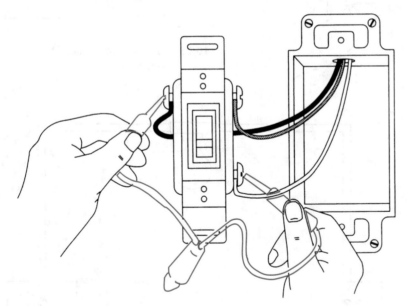

10-28 Check for voltage at a different terminal.

Four-way switches

A four-way switch is used when a light is controlled from three different locations. A four-way switch is installed between two three-way switches (FIG. 10-33). Four-way switches have four terminal screws (FIG. 10-34) and are simpler to replace than three-way switches.

To determine if the four-way switch is bad, turn off the circuit breaker and remove the switch from the box. Disconnect the wires from the

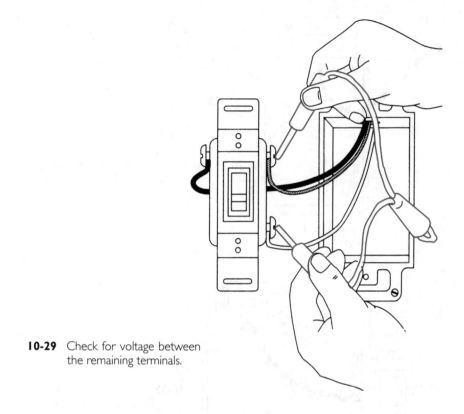

10-29 Check for voltage between the remaining terminals.

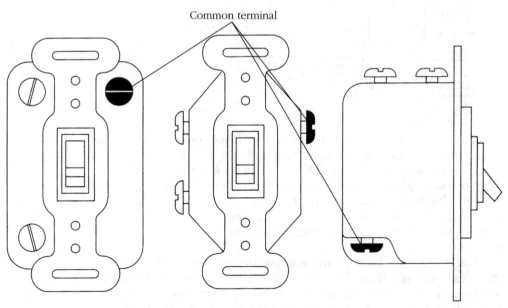

Common terminal

10-30 Locate the common terminal.

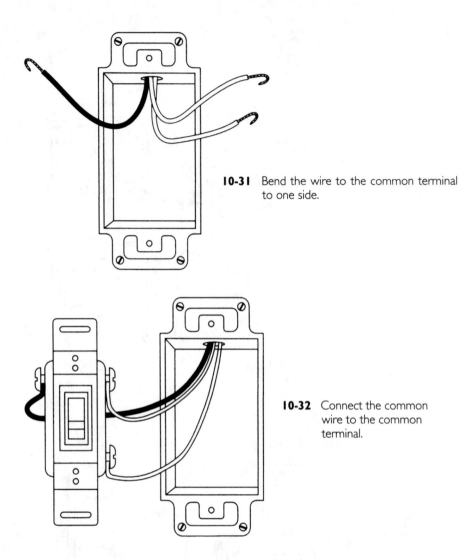

10-31 Bend the wire to the common terminal to one side.

10-32 Connect the common wire to the common terminal.

switch. Notice that, in the box, there are four wires coming from two different cables. Temporarily hook the bare end of a wire from one cable to the same colored wire of the other cable. Connect the remaining two wires the same way (FIG. 10-35). Make sure the wires are clear of the box. Turn the breaker back on and try the other three-way switches. If the other switches work correctly, the four-way switch is bad. If they only operate the light in certain positions, the four-way switch is probably good (FIG. 10-36).

Turn the breaker back off and reinstall the four-way switch. Check the three-way switches. If the four-way switch is bad, leave the breaker off and unhook the wires. Install the new switch by connecting the two wires from one of the cables to the top two terminals on the new switch. Connect either wire to either terminal. Connect the remaining two wires from

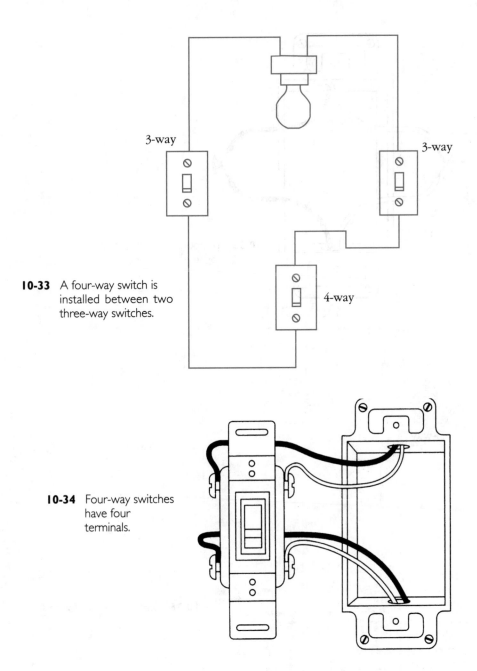

10-33 A four-way switch is installed between two three-way switches.

3-way

3-way

4-way

10-34 Four-way switches have four terminals.

the other cable to the two lower terminals on the switch (FIG. 10-37). Complete the installation and turn the breaker back on.

If it happens that the switch doesn't work, turn the breaker off, pull the switch from the box, and disconnect the two wires on one end of the switch. Interchange the wires and reconnect them to the same terminals (FIG. 10-38). The switch should work.

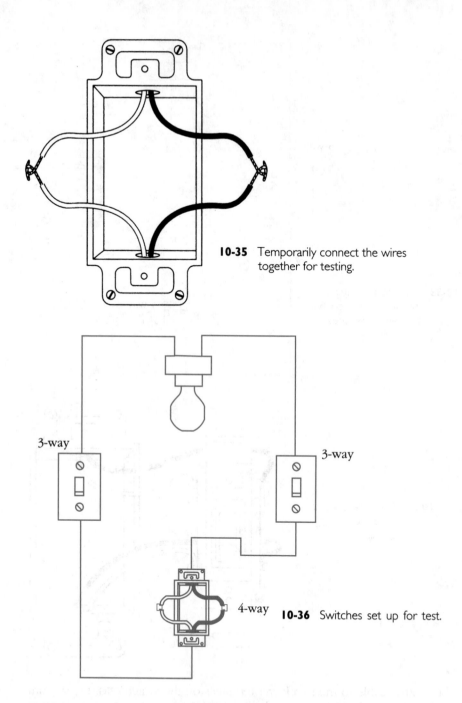

10-35 Temporarily connect the wires together for testing.

3-way

3-way

4-way   **10-36** Switches set up for test.

REPLACING DUPLEX RECEPTACLES

The most common outlet found in the home is the duplex receptacle. These seldom fail, but when they do, they can be as easily replaced as a switch. Like switches, some receptacles are made better than others, so it

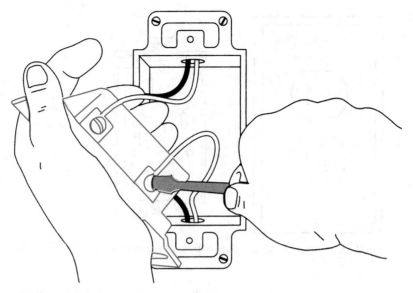

10-37 Installing a new four-way switch.

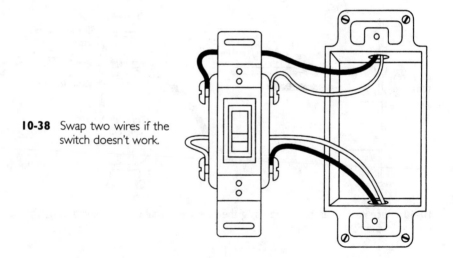

10-38 Swap two wires if the switch doesn't work.

always pays to buy quality electrical components. The only tools you'll need are a screwdriver and needlenose pliers. Always make sure the power is off to the circuit you're going to work on.

Remove the single screw holding the plate, then the plate. Loosen the two screws holding the receptacle to the box (FIG. 10-39). If the screw terminals are used, loosen them and remove the wires. If the receptacle was back-wired, insert a small screwdriver in the slot to release the wires (FIG. 10-40). Notice that the hot (black) wire was connected to the side with the brass screw and the neutral (white) wire was connected to the side with the silver-colored screw. Install the new receptacle the same way. Connect

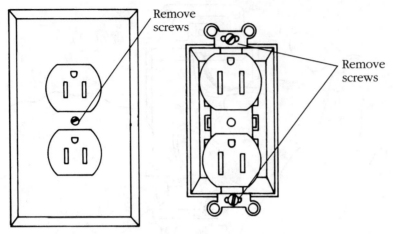

10-39 One screw holds the plate in place. Two screws mount the receptacle in the box.

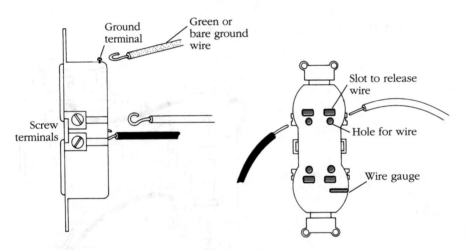

10-40 Receptacles can be wired using the screw terminal or back-wired using the holes in the back.

the ground wire to the ground terminal. The ground terminal is usually a dark-green screw.

If one end of the receptacle is controlled by a wall switch, the break-off tab between the brass screws has been removed. You need to break the tab on the new receptacle to make it identical to the one you are replacing. Grip the tab—the one on the brass side of the receptacle—with the nose of the pliers, and bend the tab back and forth. It will break after a couple of bends (FIG. 10-41). Reconnect the black wire from the switch to the top brass terminal. The white wire from the switch goes to the black source wire and the jumper wire. The jumper wire is connected to the bottom brass screw, and

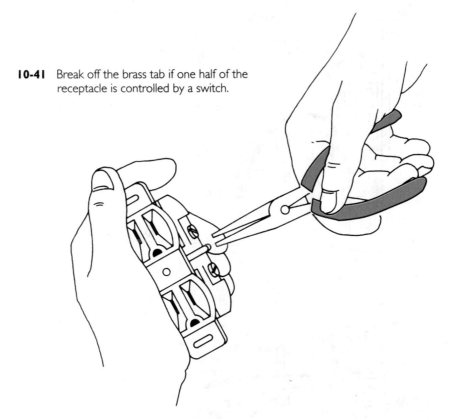

10-41 Break off the brass tab if one half of the receptacle is controlled by a switch.

the white wire from the source is connected to either one of the silver-colored terminals (FIG. 10-42).

REPLACING RANGE AND DRYER RECEPTACLES

Range and dryer receptacles very seldom cause any problems unless they have been physically damaged when an appliance was moved in or out. The procedure for replacing either one is the same. You will need a screwdriver and needlenose pliers as well as a hammer for removing the knockout in the receptacle base.

Begin by making certain the circuit is dead. Check to make sure the new receptacle is properly rated—50 amps for the range and probably 30 amps for the dryer (FIG. 10-43). Remove the single screw holding the cover, then the cover itself. Loosen the three terminal screws while noting the color code of the three wires (FIG. 10-44). Notice that the white wire is connected to the center terminal. Remove the receptacle base from the wall, loosen the cable clamp, and pull the cable from the base. To install the new receptacle, remove the appropriate knockout with a hammer or screwdriver and reverse the previous steps.

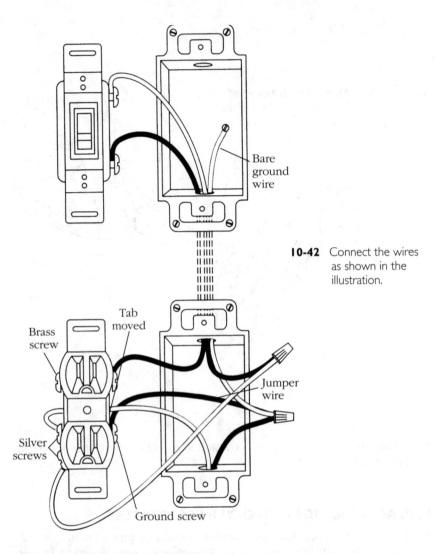

Bare
ground
wire

Tab
moved

Brass
screw

Silver
screws

Jumper
wire

Ground screw

10-42 Connect the wires as shown in the illustration.

ELECTRIC WATER HEATERS

Electric water heaters normally have two heating elements with each element controlled by a thermostat (FIG. 10-45). The thermostats are mounted on the outer wall of the tank above each element, and they sense the water temperature through the wall of the tank. To reduce energy demands, the thermostats work as a team, with one thermostat operating at a time. First, the upper thermostat turns on the top element to heat the water in the top of the tank. When the water gets hot, the thermostat turns the element off and the bottom element comes on to heat the remaining cold water at the bottom of the tank. The hot water leaves from the top of the tank and cold water enters from a tube near the bottom of the tank. The cold water is sensed by the bottom thermostat, which turns on the element to heat the water. If enough hot water is used from the tank so

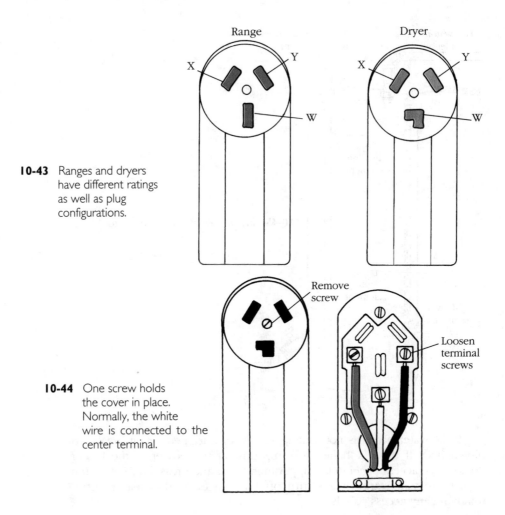

10-43 Ranges and dryers have different ratings as well as plug configurations.

10-44 One screw holds the cover in place. Normally, the white wire is connected to the center terminal.

that the water at the top drops below the setting of the upper thermostat, the top element comes back on to provide additional heat.

A safety feature is provided in case the water temperature ever reaches 210 degrees. A high-temperature cutoff on the upper thermostat turns off the power to the water heater.

Noisy plumbing when the hot water is running can be caused from too high a water temperature. Check the thermostat. If the water heater takes a long time to reheat the water after being used, there may be sediment in the tank. Turn off the electricity to the tank and the valve supplying cold water to the tank. Open one or more hot-water faucets in the house to relieve the pressure and to allow the water to drain freely. If the tank drains slowly, open more faucets. Leave the faucets open to purge the air from the tank when the tank is refilled. Next, connect a garden hose to the drain valve and drain the tank. This might take an hour or more. Open the cold-water-supply valve and run water through the empty tank until the drain water is clear.

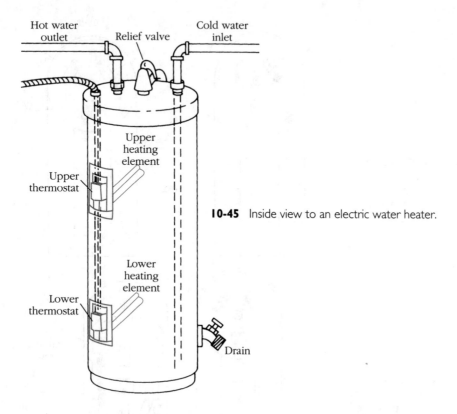

Hot water outlet
Relief valve
Cold water inlet

Upper heating element
Upper thermostat

Lower heating element
Lower thermostat

Drain

10-45 Inside view to an electric water heater.

If the water heater is not heating, check for a tripped circuit breaker or blown fuse. If that's not the problem, turn off the power to the heater. Remove the access panel to the upper thermostat and press the reset button.

Before you start, always turn off the power to the heater before removing any access panels.

Checking thermostats

Before you do anything else, turn off the circuit breaker to the water heater. Remove both access panels to the thermostats. Push the insulation back clear of the thermostats. Set the volt/ohmmeter on 250 volts ac. Hold the meter leads by the insulated covers and touch a probe to each of the top two terminal screws of the high-temp cutoff (FIG. 10-46). The two hot wires from the electrical panel are connected here. Now touch one probe to the exposed metal wall of the tank (or a mounting bolt) and the other probe to each of the terminals. In all cases, the meter should read zero volts. (If it doesn't, don't touch anything, the power is still on.) If there was no voltage present, turn the power back on and carefully repeat the test. The meter should read 220 volts across the terminals and 120 volts between each terminal and the wall of the tank. If it doesn't, you'll probably need an electrician because you're not getting power to the water heater.

If the voltage is present, check the high-temp cutoff.

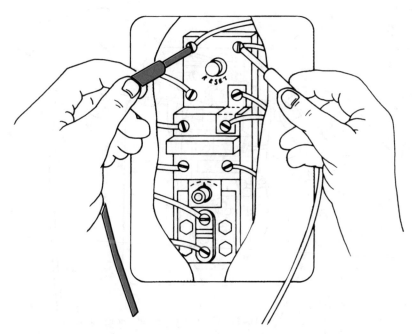

10-46 Checking for voltage to the water heater.

Checking the high-temp cutoff Turn off the breaker at the service panel. Retest for voltage at the water heater, then push in the cutoff reset button. If you hear it click, water may have seeped into one of the heating elements, causing a short. Disconnect one of the wires to the heating element (it doesn't matter which one). Set the volt/ohm on the Rx1 scale and touch a probe to each of the two screw terminals on the left of the reset button (FIG. 10-47). Repeat the test on the two terminals on the right of the button. The meter needle should move to zero each time showing a connection. If it does, check the thermostat and elements. If not, replace the high-temp cutoff.

Before disconnecting any wires, draw a diagram showing where the wires will go. Now disconnect the wires and the two metal straps going to the thermostat. Pull up on the cutoff and remove it. Push in the reset button on the new cutoff and install it in the reverse order.

Checking the top thermostat Make sure the power is off to the heater. Disconnect the wire to the top element. Use a screwdriver to turn the thermostat clockwise to its highest setting (FIG. 10-48). With the meter set on Rx1, touch one of the probes to each of the two screw terminals on the left side of the thermostat. The meter should move to zero. Now turn the thermostat to its lowest setting and repeat the tests. This time the meter should not move. If the thermostat failed the tests, replace it.

Checking the bottom thermostat Make sure the power is off to the heater. Disconnect one of the wires to the bottom element. Turn both

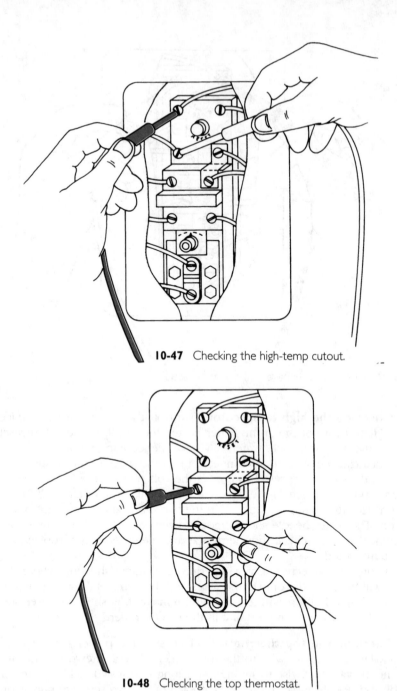

10-47 Checking the high-temp cutout.

10-48 Checking the top thermostat.

the top and bottom thermostats to their lowest setting. With the meter set on Rx1, touch a probe to each of the two terminal screws on the bottom thermostat (FIG. 10-49). The needle on the meter should not move. Now turn the bottom thermostat to its highest setting. The needle should move

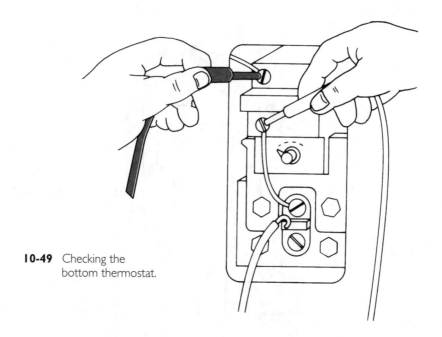

10-49 Checking the bottom thermostat.

to zero. If not, replace the bottom thermostat. If the thermostats are good, check the elements.

To replace the thermostat, make sure the power is off. Before disconnecting any wires, draw a wiring diagram. Now remove the screws connecting the two metal straps from the high-temp cutoff and the two wires from the thermostat (FIG. 10-50). Slide the high-temp cutoff up out of the way. Loosen the two bolts on the thermostat's bracket, and pull the thermostat out from the spring clips holding it to the wall of the tank. Install the new thermostat, making sure the back is flush against the wall of the heater. Tighten the mounting bolts and press the cutoff back into place. Reconnect the wires and set both thermostat to about 140 degrees.

Checking for continuity   To check a heating element for continuity, make sure the power is off to the heater at the service panel. Remove both access panels on the heater. Check again for voltage. Carefully pull the insulation back to get to the terminals. Disconnect one of the wires to each element. Set the meter to R×1 and touch a probe to each of the two screw terminals on the elements (FIG. 10-51) The needle should move to somewhere in the middle of the scale; about 10 to 20 ohms. If not, replace the element. If good, check it for a short.

Checking for shorts   To check the elements for a short—while the power is still off to the heater—adjust the meter to higher setting (R×1000 or more) and touch one probe to one of the element's mounting bolts and the other probe to one of the element's terminals (FIG. 10-52). The needle should not move at all. If it does, replace the element. There is a short circuit between the element and the tank.

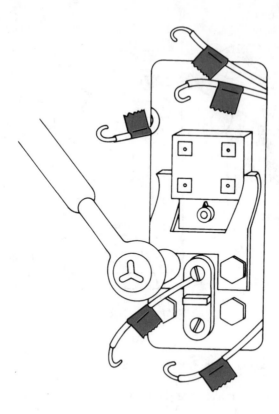

10-50 Replacing the thermostat.

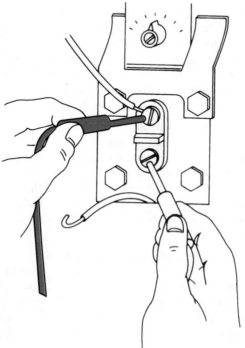

10-51 Checking for a break in the heating element.

Replacing the element Make sure the power is off to the heater. Shut off the water supply to the heater. Open several hot-water faucets at various sinks throughout the house. Use a garden hose or a bucket to drain the water from the tank. After the tank has drained, disconnect the two wires from the element. Remove the four bolts holding the element and the thermostat bracket. Now remove the element and the old gasket (FIG. 10-53). Scrape away any scale or rust with an old screwdriver so that the new gasket will make a tight seal. Install the new element and gasket. Mount the thermostat bracket and thermostat. Connect the two wires to the element. Push the insulation back in place and install the access panels. Close the drain on the tank and open the water supply to refill the tank. The tank should be full when water runs from the hot-water faucet at the sink. Turn the power back on.

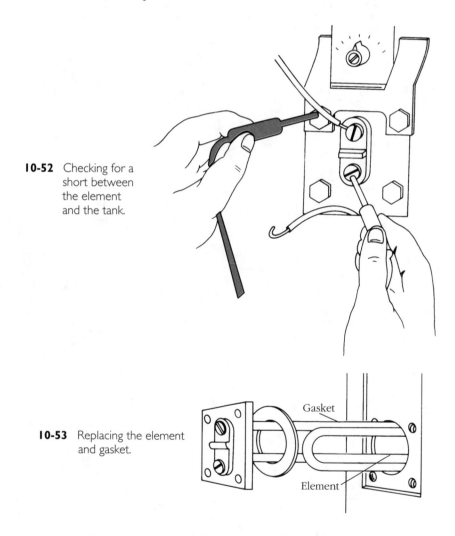

10-52 Checking for a short between the element and the tank.

10-53 Replacing the element and gasket.

Gasket

Element

Chapter **11**

Replacing light fixtures

Most light fixtures are replaced either to upgrade the old one or to remove one that has been damaged accidentally.

The tools you will need to do the job are a screwdriver, needlenose pliers, and maybe an adjustable wrench. Before removing the old fixture, be sure the circuit is dead.

REPLACING WALL-MOUNTED FIXTURES

To remove the old fixture, first turn the breaker off. Wall-mounted fixtures are usually held in place by a single, decorative cap nut. Remove any globe and light bulb. Then remove the cap nut and the fixture (FIG. 11-1). To install the new fixture, strip about ¾ inch of the insulation from each new wire. This might have already been done. Use wire nuts to connect the wires. Black wire to black wire, and white wire to white wire. Slide the fixture over the threaded nipple and hold it in place with the cap nut.

REPLACING CEILING FIXTURES

Ceiling fixtures have glass globes or shades that are held in place with thumbscrews or a cap nut (FIG. 11-2). To remove the globe with the thumbscrews, hold the globe up with one hand and loosen the thumbscrews with the other. Don't remove the screws; just back them out a little (FIG. 11-3) until the globe is free.

After the globe is removed, you will see that this fixture is held secure by two screws in a slotted opening. Just back these screws out a little (FIG. 11-4) and rotate the fixture. It should drop free of the ceiling. Remove it by disconnecting the wires (FIG. 11-5).

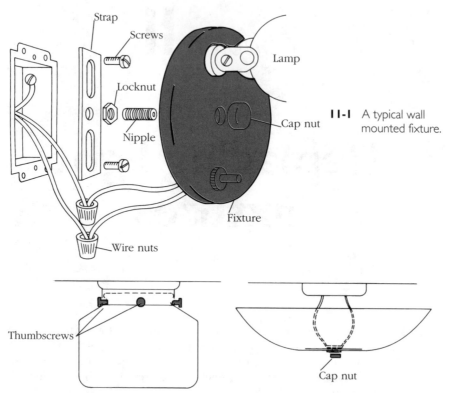

Strap

Screws

Locknut

Nipple

Lamp

Cap nut

Fixture

Wire nuts

11-1 A typical wall mounted fixture.

Thumbscrews

Cap nut

11-2 Two methods of attaching glass globes.

11-3 Remove the glass globe by loosening the thumbscrews.

11-4 Two screws in slotted openings hold the fixture in place.

11-5 Fixture removed from box.

Some ceiling fixtures with chains are held in place by two decorative cap nuts (FIG. 11-6). To remove this type, hold the weight of the fixture with one hand and remove the cap nuts with the other. With the nuts removed, lower the fixture to allow the wires to be disconnected (FIG. 11-7). Some ceiling fixtures are mounted directly to the box, while others might be mounted on a strap that is attached to the box (FIG. 11-8). Heavier fixtures might use a stud through the back of the box (FIG. 11-9). A strap is fitted over the stud and held in place by a locknut. The fixture is then attached to the strap.

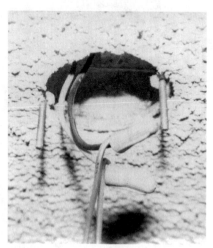

11-6 Fixture with chain held in place by two cap nuts One nut has been removed in preparing to lower the fixture.

11-7 Fixture removed and ready for wires to be disconnected.

New light fixtures come with mounting instructions for that particular fixture. Instructions will vary because some light fixtures can be quite heavy. All fixtures are designed to fit a standard outlet box, however, with wire connections being black to black and white to white with a green or bare ground wire.

REPAIRING FLUORESCENT LIGHTS

Fluorescent lights are becoming more popular in residential wiring because they cost less to operate and do not generate as much heat as a regular (incandescent) light bulb. Most work on fluorescent fixtures can be done with a screwdriver. The styles and sizes will vary, but basically there are three types available: the starter type, the rapid-start type, and the instant-start type.

The starter type (FIG. 11-10) is slow to light up. The starter must be replaced from time to time (FIG. 11-11), though it can easily be done by

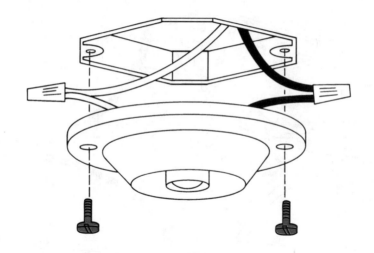

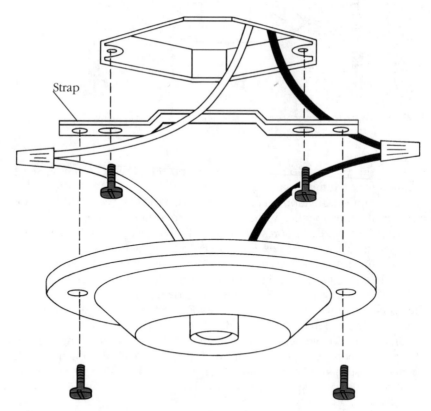

Strap

11-8 Two methods of mounting a fixture to a box.

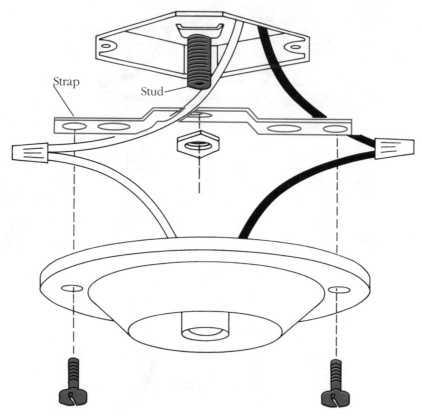

Strap

Stud

11-9 Heavier fixtures might require a stud and a strap for mounting.

pressing in and turning the starter counterclockwise. One starter is required for each tube.

The rapid-start fluorescent fixture (FIG. 11-12) is probably the most popular. It lights up quickly and does not use a starter.

The third type of fluorescent fixture is the instant-start (FIG. 11-13). It requires little maintenance except occasionally replacing a tube.

The tubes for the starter type and the rapid-start fixtures have two pins, while the tubes used in the instant-start fixtures have only one (FIG. 11-14). The ends of the tubes are marked showing the size of the tube and the type of fixture (FIG. 11-15). Discoloration at the ends of the tube often give some indication of their condition (FIG. 11-16).

Install the double-pin tubes by aligning the pins and inserting the ends of the tube about one-quarter turn. The instant-start fixture using the single-pin tube has a lamp holder with a spring-loaded socket. Press the pin on the tube into the socket far enough to allow clearance to install the other end.

To replace the ballast, make sure the power is off, remove the tubes, then remove the fixture cover to expose the wiring and ballast. Check the

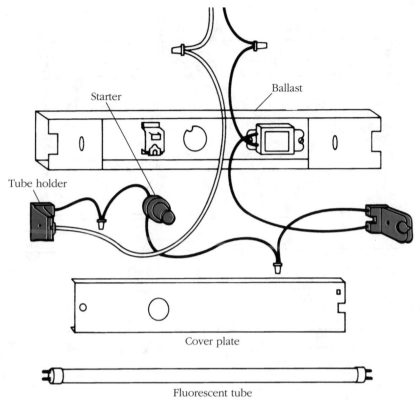

Starter

Ballast

Tube holder

Cover plate

Fluorescent tube

11-10 A starter-type fluorescent fixture.

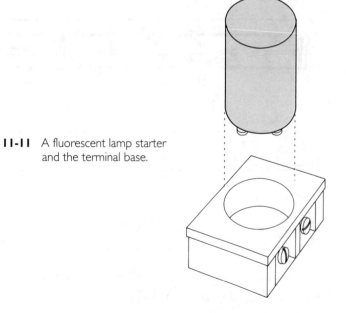

11-11 A fluorescent lamp starter
and the terminal base.

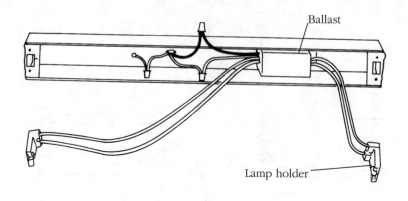

Ballast

Lamp holder

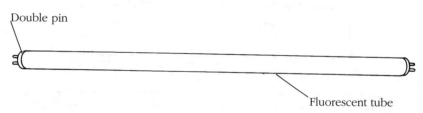

Double pin

Fluorescent tube

11-12 A rapid-start fluorescent fixture.

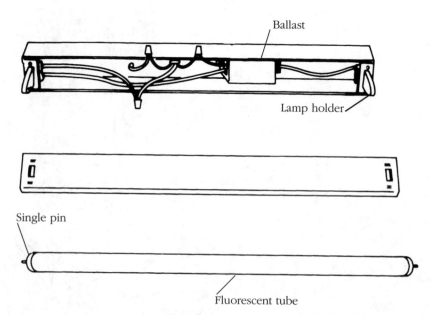

Ballast

Lamp holder

Single pin

Fluorescent tube

11-13 An instant-start fluorescent fixture.

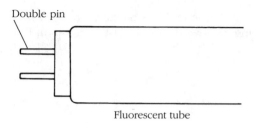

Double pin

Fluorescent tube

11-14 The two types of pin connections.

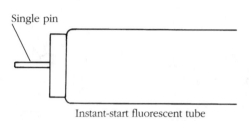

Single pin

Instant-start fluorescent tube

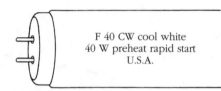

F 40 CW cool white
40 W preheat rapid start
U.S.A.

11-15 Markings on the end show the type specifications.

Blackened ends indicate faulty tube

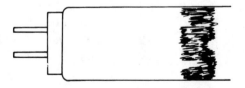

11-16 Blackened ends indicate the tube needs replacing.

Gray bands about 2 inches from ends are normal

new ballast for specifications. It should be the same as the old one and have the same number of wires. To avoid mistakes, the best way to connect the wires is to disconnect an old wire and connect it immediately to the new ballast.

Disconnect and connect one wire at a time. Next loosen the mounting, remove the old ballast, and mount the new one.

Chapter **12**

Replacing fuses, circuit breakers, & GFCIs

Any electrical circuit, no matter how small, should be protected by a fuse or circuit breaker. Fuses and circuit breakers offer protection from shock in case of a short circuit as well as protect the wiring and equipment itself. If, somehow, a hot wire comes in contact with the metal frame of an appliance, the metal frame becomes electrically "live." This hot-wire contact can damage the equipment, overheat the wire, and catch fire—or it can give someone who happens to touch the frame a severe shock.

This happens because electricity takes the easiest path to ground. In FIG. 12-1, you will see that the easiest path to the ground can sometimes be through a human body. But by connecting an equipment grounding wire to the service entrance ground, an easier path is provided, and the circuit is opened by the fuse or breaker (FIG. 12-2).

FUSES

Several types of fuses are available. The *cartridge* or *ferrule* type has ratings from 0 to 60 amps. The *knife-blade* type has ratings that run from 60 to 200 amps (FIG. 12-3).

Common plug fuses found in older homes ran in sizes including 10, 15, 20, 25, and 30 amps. When a fuse on a circuit blew frequently, it was tempting to install the next size larger. For example, a 25-amp or 30-amp fuse might be installed on a circuit using #14 wire. This size wire has an ampacity of only 15 amps. When this was done, however, the wiring inside the walls became the fuse, creating a dangerous situation, indeed.

To overcome the temptation to overfuse, a nontamperable *S fuse* was developed (FIG. 12-4). This fuse has an adapter that, once installed, cannot be removed. This adapter prevents a larger-size fuse from being installed.

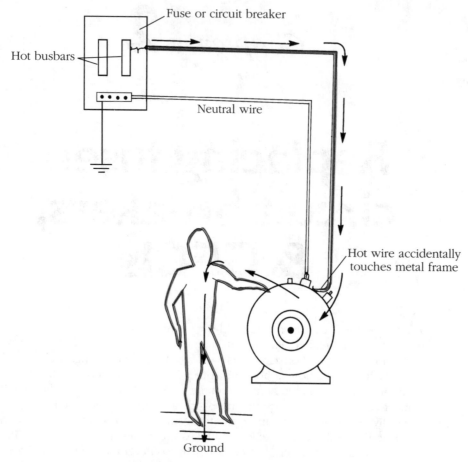

12-1 A short circuit without a grounding wire installed.

Nontamperable fuses use a short strip of an alloy with a low melting point. When the fuse is installed in a circuit, the metal strip becomes a connection in the circuit. If the current exceeds the rating of the fuse, the metal strip melts and opens the circuit. The top of the fuse has a small window that allows the metal strip to be seen (FIG. 12-5). If the fuse blows and the window is still clear, it probably was caused by an overloaded circuit where too many appliances were plugged in. But if the window was blackened by a flash, you probably have a short circuit somewhere.

REPLACING CIRCUIT BREAKERS

Residential wiring is protected by fuses or circuit breakers (FIG. 12-6). Because of their convenience, circuit breakers are preferred by most people. Inside the breaker is a tripping mechanism that will not let the breaker

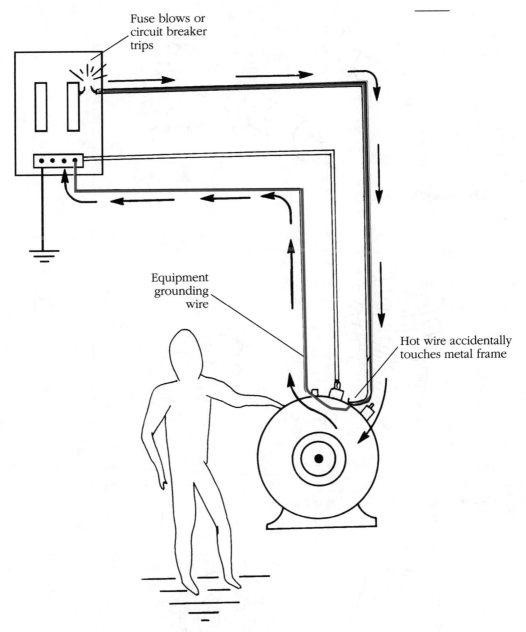

Fuse blows or
circuit breaker
trips

Equipment
grounding
wire

Hot wire accidentally
touches metal frame

12-2 A short circuit with a grounding wire installed.

be reset until cooling has taken place. Some circuit breakers have little windows that indicate when the breaker is tripped (FIG. 12-7). To reset a typical breaker, push the handle beyond the OFF position, then return it to the ON position (FIG. 12-8).

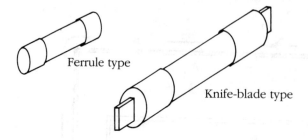

Ferrule type

Knife-blade type

12-3 Cartridge-type fuses.

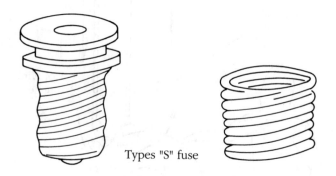

Types "S" fuse

12-4 A nontamperable "S" type fuse and adapter.

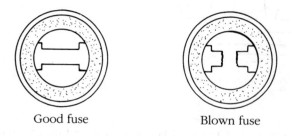

Good fuse   Blown fuse

12-5 Window in top of a fuse showing the alloy strip.

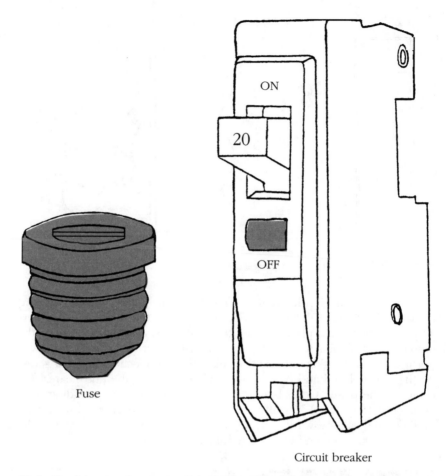

Fuse

Circuit breaker

12-6 Circuit breakers have replaced the fuse in residential wiring.

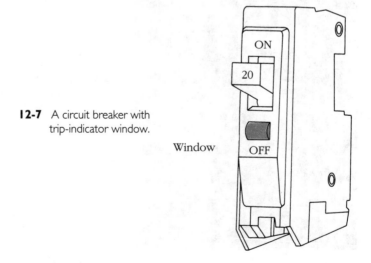

12-7 A circuit breaker with trip-indicator window.

Window

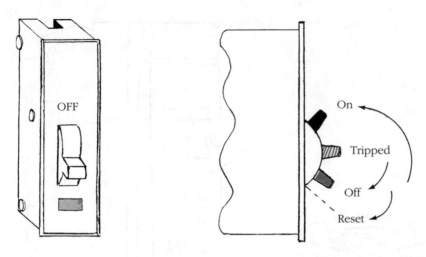

12-8 Method of resetting a tripped breaker.

To replace a circuit breaker, first turn the main breaker off and remove the cover of the panel. This will expose the individual breakers and the wires held in place by their terminal screws (FIG. 12-9). Loosen the terminal screw, then remove the wire and bend it to an out-of-the-way position. Some breakers are held in place by a projection under a screw on the terminal end and by the electrical contacts on the end nearest the center of the panel. Other brands have metal projections extending from both ends (FIG. 12-10). Usually, you can remove the circuit breaker by pulling the breaker toward you. Begin at the end toward the center of the panel and rotate the end with the terminal screw out last.

12-9 Service entrance panel with the cover removed to show terminal screws on circuit breakers.

12-10 A circuit breaker. The mounting clip on the right, and the contact plug-in that makes electrical connection to the hot busbar is on the end to the left.

REPLACING GFCIs

Ground fault circuit interrupters (GFCIs) are now required on all outdoor receptacles and any receptacle serving a swimming area. In new construction, GFCIs are required in bathrooms and in at least one location in a garage.

All ground fault circuit interrupters have a test button, which should be used to test the unit about once a month. If the bottom is pushed and the GFCI does not trip, replace the GFCI. Usually all you need is a screwdriver and needlenose pliers to do the job.

Receptacle GFCIs

Replace the receptacle-type GFCI as you would any receptacle. Make sure the power is off, then remove the old GFCI. The new one will have a TEST and RESET button. Wire the new GFCI according to standard practice—black wires to the brass screws and white wires to the silver-colored screws (FIG. 12-11). Connect the ground wire to the green ground terminal. If the receptacle is at the end of the run (no other receptacles connected downstream), do not use the bottom terminals.

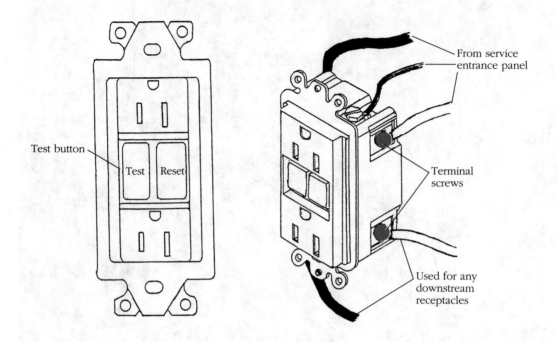

12-11 A receptacle GFCI and wire connections.

Circuit-breaker GFCIs

The circuit-breaker GFCI takes a little more work to replace. It protects an entire circuit and is installed in the service entrance panel. The main circuit breaker should be turned off before you begin work. You probably will need a screwdriver, needlenose pliers, and wire strippers.

The new GFCI breaker will have a TEST button, but it also will have a curly white wire (FIG. 12-12). Begin by turning off the main circuit breaker and removing the panel cover. Next, remove the old breaker and disconnect the wires. The GFCI breaker terminals are marked NEUTRAL and POWER. Connect the white (neutral) wire to the terminal marked NEUTRAL and the hot (black) wire to the terminal marked POWER (FIG. 12-13). The breaker can now be installed in the old breaker's slot. Next, connect the white curly wire to the neutral busbar (FIG. 12-14). The completed installation should be similar to the one shown in FIG. 12-15.

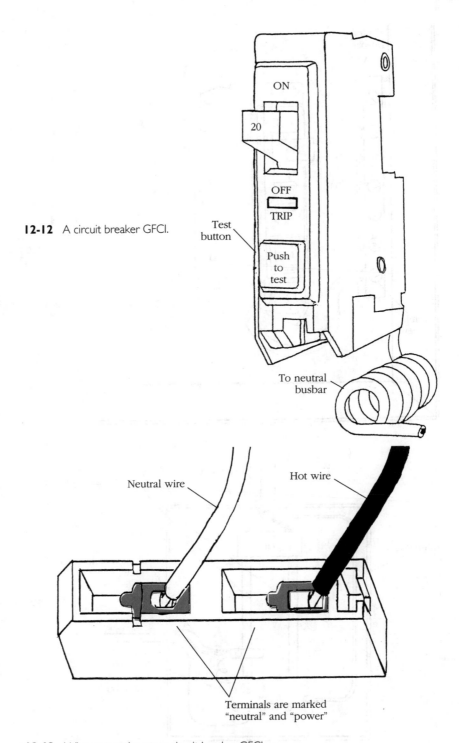

12-12 A circuit breaker GFCI.

ON

20

OFF
TRIP

Test
button

Push
to
test

To neutral
busbar

Neutral wire

Hot wire

Terminals are marked
"neutral" and "power"

12-13 Wire connections to a circuit breaker GFCI.

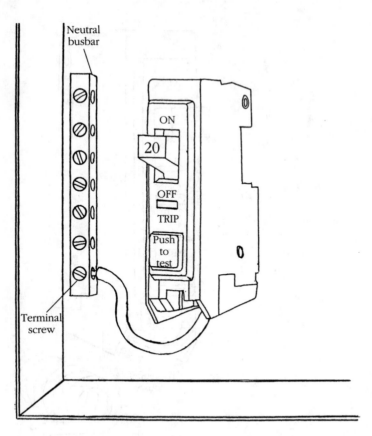

12-14 A circuit breaker GFCI has a white wire connected to the neutral busbar.

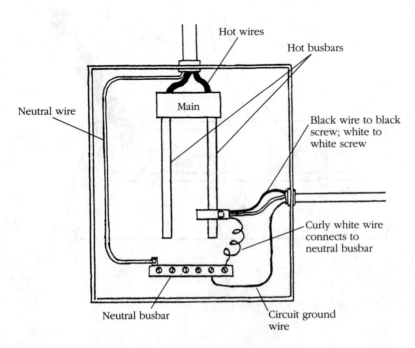

12-15 Typical installation of a circuit breaker GFCI.

Glossary

alternating current Current that regularly reverses its direction, flowing first in one direction, then the other. Abbreviated ac.

ampacity The amount of current, expressed in amps, that an electrical conductor can carry continuously without exceeding its temperature rating.

ampere A unit used in measuring electrical current. Abbreviated amp. It is based on the number of electrons flowing past a given point in one second. It can be determined by the Ohm's Law formula: I (current) = watts ÷ volts.

AWG (American Wire Gauge) The adopted standard of wire sizes such as No. 12 wire, No. 14 wire, etc. The larger the number of the wire, the smaller the size of the wire. A No. 14 wire is smaller than a No. 12 wire.

bonding The permanent joining of metal components of an electrical system to form a continuous, conductive path capable of handling safely any current likely to flow.

bonding jumper A reliable conductor between metal parts where the parts might be temporarily separated at some time.

bonding jumper, main The connection between the neutral bar in the service entrance panel and the panel.

branch circuit Any one of a number of separate circuits distributing electricity from an overcurrent protection device.

busbar The solid metal conductor in a distribution panel that provides a common connection between the main service and the branch circuits.

cable A stranded conductor or a group of individual conductors insulated from each other.

cable, entrance A heavy cable used to supply electrical service from the main line to a building.

cable, nonmetallic sheathed Two or more insulated wires assembled inside a plastic sheath. Type NM is used in dry locations. Type NMC can be used in both dry or damp locations.

circuit An electrical conductor forming a continuous path, allowing current to flow from a power source through some device using electricity and back to the source.

circuit breaker A safety switch installed in a circuit that automatically interrupts the flow of electricity if the current exceeds a predetermined amount. Once tripped, a circuit breaker can be manually reset.

conductor The trade name for an electric wire or busbar capable of carrying electricity.

continuity The state of having a continuous electrical path.

current The transfer of electrical energy caused by electrons traveling along a conductor. Abbreviated I.

cycle One complete reversal of alternating current where a forward flow (positive) is followed by a backward flow (negative). The standard rate in the United States is 60 cycles per second, now called 60 Hertz.

duplex receptacle A receptacle providing electrical connections for two plugs.

fuse A safety device installed in a circuit designed to interrupt the flow of electricity should the current exceed a predetermined amount. Fuses cannot be reused.

ground A conducting connection between earth and an electrical circuit or equipment, whether intentional or accidental.

ground clamps Metal clamping devices used to connect a wire to earth ground, often through a ground rod or water pipe.

ground fault circuit interrupter A safety device installed in a circuit designed to protect people by detecting very small currents and interrupting the flow. Abbreviated GFCI.

grounding electrode conductor The conductor from the neutral busbar in the service entrance panel to the established ground.

hot busbars The solid metal bars in service entrance panels and subpanel where the power source is connected. Circuit breakers or fuses mounted on these busbars deliver power to individual branch circuits.

hot wires The ungrounded current-carrying conductor of an electrical circuit. Normally, it is identified by black or red insulation, but it can be any color except white, gray, or green.

insulation A nonconducting material used to cover wires and components to remove shock hazards and to prevent short circuits.

junction box A box containing only the connections or splices of several wires.

kilowatt A unit of electrical power measured as 1000 watts. Abbreviated kw.

kilowatt-hour The unit used in metering electricity. One kilowatt-hour is 1000 watts used in one hour or the equivalent. For example, it might be 500 watts used for two hours. Abbreviated kwh.

knockout Round, partially punched-out openings installed by the manufacturer in panels and junction boxes that allows the opening to be knocked out with a screwdriver.

low-voltage wiring A method of wiring where a lower voltage is used to supply electricity for such purposes as doorbells, thermostats, and some outdoor lighting.

neutral busbar The solid metal bar in a service entrance panel or subpanel used as a common terminal to connect all of the neutral wires. The neutral busbar in the service entrance panel is bonded to the panel as well as being directly connected with earth ground. The neutral busbar in the subpanel is used to connect the neutral wires, but it is isolated from the panel and is not grounded.

neutral wires The grounded conductor that provides the return path to the source to complete the circuit. Neutral wires must never be interrupted by circuit breakers or fuses. Neutral wires are identified by white or gray insulation.

ohm The unit for measuring electrical resistance.

ohmmeter The instrument used in measuring electrical resistance.

outlet A point in a wiring system where current might be taken to supply electrical equipment.

overcurrent protection device A fuse or circuit breaker that automatically interrupts the flow of electricity in the event the current exceeds a predetermined amount.

overload A situation where an electrical circuit is attempting to carry more current than it can safely handle.

pigtail A single wire extending from a connection of two or more wires.

receptacle A connecting device designed to accept plugs.

resistance The property in an electrical conductor or circuit that restricts the flow of current. It is measured in units called ohms.

service drop The overhead wires from the utility pole that delivers the electricity to the building.

service entrance panel The main power cabinet containing the main breaker and circuit breakers distributing electricity throughout the residence.

service lateral Underground wires providing electrical service to the building.

short circuit A completed, very low-resistance circuit, where two bare hot wires come in contact or a bare hot wire touches a bare ground wire or grounded component.

source The point supplying the electrical power. It might be a battery, generator, or the service entrance panel.

switch A device used to close or open a circuit, allowing current to flow or not flow, respectively.

terminal A point used to make electrical connections.

thermostat A device used to control temperatures to a predetermined level.

transformer A device used to transfer electrical energy from one circuit to another using electromagnetic induction. Two types commonly found are the step-up and step-down transformers. The step-up transformer is used where a higher output voltage than the input voltage is desired. A step-down transformer is used for circuits where a lower output voltage than the input voltage is needed, such as with doorbells and thermostats.

upstream/downstream The location of a point in a circuit relative to the power source. Upstream refers to the part of the circuit between the point and the power source. Downstream refers to the part of the circuit from the point to the remaining part of the circuit going away from the power source.

volt The unit used in measuring electrical pressure. Abbreviated V or E.

voltage The electrical pressure. measured in volts, at which a circuit operates. Voltage can be determined by the Ohm's Law formula: voltage = watts ÷ current.

voltage drop An electrical term describing the loss of voltage that occurs when the wires are not of sufficient size to carry the amount of current flowing.

voltmeter A meter used to measure voltage.

watt The unit used in measuring electrical power. Abbreviated W. The amount of power in a circuit can be determined by the Ohm's Law formula: watts = current × voltage, or $W = I \times V$.

Index